# SpringerBriefs in Computer Science

*Series Editors*

Stan Zdonik
Peng Ning
Shashi Shekhar
Jonathan Katz
Xindong Wu
Lakhmi C. Jain
David Padua
Xuemin Shen
Borko Furht
V S Subrahmanian

For further volumes:
http://www.springer.com/series/10028

Kerstin Denecke

# Event-Driven Surveillance

## Possibilities and Challenges

 Springer

Kerstin Denecke
Forschungszentrum L3S
Hannover
Germany

ISSN 2191-5768
ISBN 978-3-642-28134-1
DOI 10.1007/978-3-642-28135-8
Springer Heidelberg New York Dordrecht London

e-ISSN 2191-5776
e-ISBN 978-3-642-28135-8

Library of Congress Control Number: 2012931408
ACM Codes: H.3, I.2, I.5, I.7

Printed on acid-free paper

Springer is part of Springer Science+Business Media (www.springer.com)

*I am surveilled therefore I am.*

# Preface

The Web has become a rich source of personal information in the recent years. People are twittering, blogging and chatting online. Current feelings, experiences or latest news are posted everywhere. Face-to-face communication seems to be replaced by the new communication tools. Since this new communication style provides insights into thoughts, behaviour and others, it offers the opportunity to make use of the data in other contexts as in critical situations. For instance, first hints to disease outbreaks or political changes could be identified from this data. Further, companies discover the possibilities of social Web communication. They collect knowledge for adapting marketing strategies from the Web chatter.

The author's goal in this book is to show the opportunities for real-world surveillance provided by the new communication means. The existing technological possibilities for online surveillance are the focus. The author provides a general guideline on how to choose technologies for developing a surveillance system depending on the user, the available content and the scenario of use. Furthermore, open research and development issues are presented.

This book is about surveillance systems, that is, about systems that monitor subjects or their behaviour. I concentrated on systems and approaches that exploit unstructured data for surveillance purposes. While this topic is prevalent in the field of biology and medicine, other domains only started to become aware of the necessity of monitoring tools. In particular, with the immense growth of the Web, this topic becomes extremely important for other domains. There are books available in the market dealing with biosurveillance. Then, there are books available on event detection in the business domain. With this book I aim to provide a comprehensive book that provides insights into the practical and technological opportunities of surveillance across domains to guide developers and decision makers to the existing tools and methods, but without focussing too much on specific technologies.

The present version of the book represents the results of about two years of work. The book reflects the research interest of the author, going into depth at some point while providing an overview only on other points. I 'however' think this book provides a good balance and it provides a variety of readers with interesting and stimulating contents.

Many people contributed in various ways to the publication of this book. This book is inspired by the M-Eco project (http://www.meco-project.eu) that was funded by the European Commission. Parts of the book resulted from the author's project activities and discussions with colleagues in the context of the project. I give thanks for the interesting collaborations and discussions with all persons working together in the M-Eco consortium. Further, I thank Ralf Gerstner and the reviewers at Springer for their belief in the value of this book and the support in production.

Finally, I would like to thank my friends for encouraging me during the writing of this book.

Hannover, December 2011                                              Kerstin Denecke

# Contents

# Chapter 1
# Introduction

We are living in an era where homes and environments become smart through technology. They are giving people the opportunity to monitor and react to changes in the environment very quickly. There have been examples where monitoring Twitter messages prevented from terrorist attacks. In such critical situations, being informed early about changes in situations or in opinions can be crucial for reacting in time. This holds true for multiple domains. Rumours about political changes can provide hints to upcoming changes in the financial market. A change in the opinion of a population towards its government might be a first sign of a revolution. Just remember the political changes in North Africa in 2011 where people connected via Twitter. Monitoring such chatting could provide insights into the situation and would allow the various involved parties to react appropriately. Another example is the early detection of disease activity which is the objective of disease surveillance: Hints to potential outbreaks of infectious diseases or to expected food contamination need to be analysed. For all this holds true: Reactions can be taken only in time when decision makers are aware of a problem at an early stage.

Surveillance or early warning systems enable this detection of changes and support humans in getting information on changing situations. They collect data from multiple sensors, analyse the data and provide a summary or interpretation to the user. The variety of data that could be considered for surveillance is immense, ranging from sensor-measured values to collected counts and information extracted from natural language documents. All this information need to be analysed, interpreted and summarised for the purpose of monitoring situations or objects.

Without information technology we would be unable to process all the data available. In our information society data is stored everywhere, often without having in mind the potential re-use of those data for surveillance from the beginning and often without the knowledge of the single individual. Several years ago, technologies were missing to process these huge amounts of data. In particular with the increased availability of Web data, new data sources for surveillance came up and are posing new challenges for surveillance-supporting technologies. Real-time processing of incoming streams of unstructured data—this is the problem to deal with in

K. Denecke, *Event-Driven Surveillance*, SpringerBriefs in Computer Science,
DOI: 10.1007/978-3-642-28135-8_1, © The Author(s) 2012

future surveillance. Given the promising potential for surveillance applications and assuming that the challenges of real-time monitoring and event detection can be tackled adequately, it is worthwhile to investigate which techniques are appropriate for this purpose.

The objective of this book is to introduce the multiple possibilities and facets of surveillance and its applications. Thus, this book will first introduce the task of surveillance and an overview on surveillance in various domains will be given. Then, the various information sources that are available and could already be used by surveillance systems will be summarised. In the main part of the book, the focus is on unstructured data as source for surveillance. An overview on existing methods and methods to be developed in order to process this kind of data with respect to surveillance will be given. In this way, the reader will get information on relevant methods that can be exploited when building surveillance systems. As an example application, disease surveillance using Web 2.0 will be introduced with corresponding methods and challenges to be addressed. The book will finish with remarks on new possibilities for surveillance gained by the development of the Internet and of mobile communication in the last years and with an outline of future challenges. These open issues and future challenges are intended to provide a starting point for researchers to follow-up with the research in this interesting area.

# Chapter 2
# The Task of Surveillance

Intuitively, the term *surveillance* produces rather negative feelings. We start remembering the book of George Orwell "1984", where a society is described that monitors the complete life of its members. However, doing surveillance is often crucial to survive: Even zebras on the savannah are doing surveillance when they sense and respond to the opportunity of a water hole or the threat of a lion. Surveillance is intended to monitor persons, situations or processes often following the objective, to avoid harm from population groups or objects or just to be able to react to situational changes. In this chapter, the task of surveillance with its facets and application areas is described.

## 2.1 A Motivating Example

As starting point, we consider an example from the area of disease surveillance. Global warming, globalisation, increased mobility of persons and air traffic contribute to the increased speed of spread of infectious diseases. Their effects to public health can be immense. More and more people can be infected and even die. Just consider the outbreak caused by avian flu in 2002 where thousands of people were infected. Without knowing about the origin of infections and numbers of infected persons, nobody can take precautions or react in time. This is the reason why disease surveillance tries to identify hints to public health threats as early as possible. Its objective is to prevent harm caused by serious infections from the population. The following example is intended to show (1) what surveillance can be, and (2) how information technology can support surveillance processes.

*The 2010 European Game of Nations has begun. Officials have anticipated that over 2.5 million people from across the world will descend on Europe to participate in this bi-centennial event. During a one month period, over seventy games in twelve different members states are scheduled. Concerned by recent security threats and the emergence of avian influenza in Europe, public health officials have heightened their awareness for a good infectious disease surveillance strategy during the games.*

K. Denecke, *Event-Driven Surveillance*, SpringerBriefs in Computer Science, DOI: 10.1007/978-3-642-28135-8_2, © The Author(s) 2012

*Maria, working as epidemiologist at the national health organisation in Germany, as well as key personnel in other member states, is fully aware that mass gathering and population movement increases the risk and spread of infectious diseases across borders. She specifies in the surveillance system she is using, in which information she is interested, e.g. which diseases or symptoms she would like to monitor and in which area. Among others, she specifies symptoms such as* fever, headache, diarrhoea, nausea.

*The surveillance system searches various sources for relevant information. As a daily routine it checks its indexed blogs tracking and detecting topics, locations, and diseases that have been extracted. For interpretation purposes, the surveillance system relates these indicators collected from various sources in time and space. By checking for passed thresholds, determining quick and unpredicted changes with significant differences, and by studying trends, anomalies are detected and alerts are generated.*

*One day within the games, the surveillance system reveals high (Web) activities centred around the topic "European Game of Nations" and "meningococcal disease". A check against the knowledge database of severity indicators triggers the system that there is an event of public health relevance. Since the location detected is in Germany and Maria's user profile specified her interest in symptoms that are also characteristic for the* meningococcal disease, *Maria receives an alert. The information that led to the signal is presented to her. Maria can drill down into the actual documents, audio transcripts, etc. to validate the generated signal.*

*Additionally, Maria can directly check the surveillance database for reports of cases of a reportable infectious disease as well as for outbreak reports. Based on this information, she can decide whether or not the signal is indeed an event of public health relevance. Since she can not find a related outbreak report in the database and the associated information sounds serious, Maria finally concludes that the signal points to something previously undetected. She informs her colleagues and together with the ministry of health, they decide for actions to be taken.*

## 2.2 Definitions and Characteristics of Surveillance

### 2.2.1 Surveillance Task and Objectives

**Surveillance** is a systematic, continuous monitoring. More generally spoken, surveillance targets at knowing what is going on. The subject under surveillance are persons, objects, their actions, situations or changes that concern them. Often, the changes observed have negative consequences (e.g., dieing people due to a disease outbreak), but also positive changes might happen (e.g., positive market developments). The main objective of surveillance is to gather situational awareness. Being aware of a situation means that we know what is going on, what led to the situation and what risk is associated to it.

Surveillance is performed by everyone, be it consciously or not: animals, people and companies survive and thrive based on their ability to act quickly on opportunities and threats. A Wall Street trader monitors arbitrage opportunities in sub-second response times. Banks track credit card usage to stop fraudulent charges as they occur. You are monitoring the traffic on the radio, when you go to work to avoid that you end up in a bad traffic jam. Health authorities monitor the health status of the population etc. The examples are manifold, but the objectives of doing surveillance are similar:

- identifying critical situations or changes in situations,
- identifying high-risk groups,
- identifying the most serious or the most prevalent conditions,
- monitoring the trends of these conditions and the impact of interventions,
- reacting in time,
- avoiding serious problems,
- supporting decision making.

Critical situations or changes of the state of an object are referred to as events. More specifically, an event can be everything, e.g., an action, a process, a decision, that happens or that is expected to happen [9]. It may be a threat, an opportunity or any other kind of situation that requires timely responses. Events can occur in several levels: technical events (e.g. the change of a temperature at a machine), system events from an IT infrastructure (e.g. http request) or events that reflect some complex process (e.g. crossing a threshold of persons infected with measles, traffic jam). A significant state change is a change in the "real" state that deviates from the "expected" state and where the deviation is significant enough to cause a change in plans or to require some kind of reaction.

Another relevant term in the context of surveillance is the term **indicator**, which refers to a quantity that is based on clearly defined events. It is a measurement that, when compared to either a standard or the desired level of achievement, provides information regarding an outcome or a management process. Given the broad range of information sources for surveillance (see Chap. 3) indicators can be different things. Mentions of specific terms can be considered as indicators. The frequency of the search term *flu* is, for instance, an indicator when monitoring influenza using Web search queries. The number of reported cases of a disease or an infection according to a case definition are considered as indicators in disease surveillance.

A **signal** is generated by a surveillance or alerting system from indicators detected on observed data sources whenever they reflect abnormal behaviour. It can therefore be considered as hint to a possible event. The threshold might be fixed by a user or be externally defined.

## 2.2.2 Surveillance Process

A surveillance process comprises several steps. From the data perspective, this process is as follows: at the beginning, there is a real world **event**. An event results

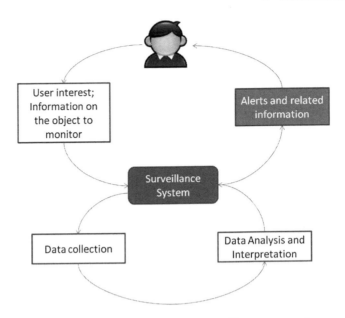

**Fig. 2.1** The surveillance process: Interaction between surveillance system and user

in changes in indicators. The analysis of indicator data can result in a signal when a real world event happened.

From the human analyst perspective, a surveillance process comprises five essential steps:

1. Definition of essential data to collect,
2. Systematic collection of data,
3. Organisation and analysis of data,
4. Implementation of interventions based on the data; and
5. Re-evaluation of interventions.

Some of these steps can be realised with computers, e.g., the collection of data. Others can only be performed by humans (e.g., reacting to an identified disease outbreak). Nevertheless, **surveillance systems** aim at supporting this process. In particular, the identification of risks and related information to help the user in information gathering and decision making is supported (see Fig. 2.1). When building a surveillance system or setting up a surveillance process, it needs to be specified, what will be monitored, and what are the expectations of normal behaviour of the object under monitoring. Further, it needs to be clarified, what a significant deviation of reality from this expectation means and which reaction needs to be performed when a change happens. All these specifications can only be made by human analysts.

After specifying these surroundings of the system, the characteristics of essential data to be collected need to be specified. The data need to reflect the current situation or behaviour of a person or object under monitoring. Then, the data need to be collected continuously from respective sensors. In regular time intervals updates from the data need to be collected, analysed and checked for critical situations. This continuous monitoring is crucial to identify changes in time. The time interval depends on the subject under surveillance and on the monitored source. Thus, when building a surveillance system, the question need to be answered in which time intervals changes are expected. Accordingly, the data update intervals can be set. In Chap. 3 possible data sources that could be considered for data collection are described.

Next, the data need to be analysed with respect to a surveillance task. The **surveillance task** is the focus of the surveillance. In disease surveillance for example, the surveillance task is monitoring behaviours or situations that may indicate a change in population health. In airport surveillance, the task is to monitor for behaviour that may indicate a threat to the airport (e.g., terrorist activities). Given the immense volumes of data available, automatic support by surveillance systems for analysing the data is becoming more and more crucial. The collected data is analysed with respect to some given criteria, e.g. some thresholds crossed, the polarity of an opinion reverted. A variety of analytical processes can be applied to convert data into information that can be used for surveillance. The data analysis might result in a signal to an event which is also the output of the surveillance system. It produces alerts or signals that are presented to the user. The overall surveillance process is summarised in Fig. 2.1.

### 2.2.3 Summary

Surveillance is the systematic, continuous monitoring of persons, their behaviour, actions, situations or processes. Its *objective* is to avoid harm, react appropriately to situational changes or critical situations, to observe ongoing activities or individuals. Data about the subject under surveillance is *collected* through sensors, *analysed and interpreted*. This process can be done either manually by humans or automatically by surveillance systems. Surveillance systems present results pointing to changes or events to the user. In the Fig. 2.2, the terms relevant within surveillance are summarised in a hierarchically ordered terminology tree.

## 2.3 Application Areas

As stated before, surveillance is of relevance in multiple domains including politics, finances, public health or traffic management. In this section, an overview on possible surveillance applications is given. Again, the focus is on surveillance systems that base mainly upon unstructured data (i.e. natural language documents).

Surveillance
    Concepts
        Event
        Indicator
        Signal
    Technology
        Surveillance system
        Content collection
        Data analysis
        Signal generation
        Result presentation
    Data
        Structured
        Unstructured
    Surveillance Process
        Identification
        Analysis
        Evaluation

**Fig. 2.2**  A hierarchy of terms related to surveillance

## 2.3.1 Surveillance in Politics

Surveillance in politics is many-fold and comprises among others media and opinion monitoring or terrorism surveillance. Monitoring for terrorist activity targets at preventing future terrorist attacks and minimising the damage and recovery from attacks that occur. Surveillance systems running in that context issue timely critical alerts and can—in this way—help to prevent attacks or crimes [2, 19]. Information technology is used by terrorism surveillance systems to support knowledge discovery by collecting, processing, and analysing terrorism- and crime-related data.

Beyond, mining and monitoring opinions is of interest in politics. The target under monitoring are, in that case, opinions expressed and statements made. For instance, politicians can monitor elector's suggestions or claims, or the perception they might have about leaders' statements. New data sources for political surveillance became available with the social Web. Social media data can provide real time and often first hand information on political situations. During the Iranian election Twitter was used as a powerful tool to get news out of the country. The cartoon by D. Marsden in Fig. 2.3 shows this development. As New York Times[1] wrote already in October 2010 about the growing use of sentiment analysis technology in politics, consultants and pollsters are now using it to analyse social media chatter and take the pulses of voters. Among others, Twitter was used to monitor the U.S. presidential debate in 2008 [20] and to predict political election results in that period. It turned out that Tweeters tended to favour Obama over McCain, and Obama really won the election.

Systems such as Twitter Sentiment[2] allow to search the sentiment expressed with respect to a person, brand, product, or topic. They provide statistics on the kind of

---

[1] http://www.nytimes.com/2010/11/01/technology/01sentiment.html?

[2] http://twittersentiment.appspot.com

**Fig. 2.3** In June 2009, during the Iranian Election Twitter transformed the way people convey news. Cartoon of D. Marsden, 2009

opinions in terms of negative and positive and can show the sentiment development over time. In TwitterSentiment it could be seen on May 4, 2011, that the number of positive and negative tweets on "Obama" increases since this was the day when Bin Laden was killed.

Monitoring opinions in politics is crucial for several reasons. Politicians can learn about what people are saying in traditional and social media about themselves and their opponent. They can identify relevant topics in these online discussions. It allows not only to react to news articles and TV, but also to address the opinions of the population expressed in social media. Beyond, current (political) activities in one's own or neighbouring countries can be assessed by monitoring the media.

Existing means for doing political surveillance include media monitoring services for online news, social media data or multimedia data as well as sentiment analysis tools. The European Media Monitor[3] is a tool provided by the Joint Research Centre that monitors news portals, classifies the articles, analyses the news texts, aggregates the extracted information and issues alerts. It allows for detecting trends in stories over time.

SAIL Labs[4] develops technologies to mine multimedia data and text. The technologies support commercial applications (e.g., news monitoring, broadcast news transcription and indexing) as well as government applications (e.g., public opinion monitoring, surveillance and intelligence gathering). The software allows, for instance, to monitor dozen (or hundreds) of TV & Radio channels or even certain keywords within these channels live at the same time.

---

[3] http://emm.newsbrief.eu/overview.html

[4] http://www.sail-technology.com/

In summary, surveillance in politics targets at monitoring public opinions, topics of public interest or terrorist activity. Available technologies allow to monitor media channels with respect to some political event or monitor the opinion of the public.

## 2.3.2 Surveillance in Business and Finances

*Environmental scanning* is the acquisition and use of information on events, trends and relationships in the external environment of an organisation [74]. It helps an organisation to adapt to changes in the environment and to react appropriately with the objective of securing or improving its future market position. Other objectives of such monitoring are market and marketing monitoring or studying the success of campaigns. In medium-size enterprises, media monitoring targets at the identification of customer opinions, analysis of strengths and weaknesses or monitoring of competitors. The idea is to scan the environment, detect changes and hotspots by monitoring financial reports and online media.

As in the domain of politics, surveillance of sentiments plays an important role in the business domain. Businesses were quick to embrace sentiment analysis to monitor the performance of their brands. Software tools scrape online messages for mentions of certain products, then determine whether the language used expresses positive or negative feelings. For instance, traders are interested in the reaction of politicians or the public to political changes. Interpretations of such reactions as signs to an upcoming inflation are possible and necessary to react appropriately. Several researchers showed that from financial internet message boards and online social media tools, indicators of financial phenomena can be identified [17, 68].

Surveillance is further of interest within the processes of gathering competitive intelligence. These processes target at strengthening the market position of a good or service by comparing and contrasting the product with similar goods and services that are currently available. In order to gather enough information and to monitor the market the following sources are of interest:

- Potential buyers. What are their current opinions about a company's products? What is the current demand?
- Competitors. Does a competitor releases a new product? What are the competitor arguments for the new product? Are there new products coming up?

A business surveillance system collects information that provides answers to such questions. For a list of competitors, it could provide alerts when information on the release of a new product becomes available. Another surveillance task could be the monitoring of opinions expressed towards the business, a company or a product. Within business intelligence such monitoring takes already place. However, it is often done manually by skimming through press releases of companies or other information sources.

In addition, companies started exploiting the benefits of social media monitoring for their business purposes. In this case, surveillance systems search various social media sources and present the results to the user by means of graphs and result lists. Automatic monitoring currently provides quantitative results only, i.e. the number of positive or negative opinions is provided or at least the number of persons that post something on a product. A qualitative analysis of the textual content is still part of ongoing research activities. Such analysis could provide information about quality of services and products as recognised by users or customers. This information can be exploited for product adaption and improvement. Even ideas for new products can probably be identified.

## 2.3.3 Disaster Surveillance

A disaster is a serious disruption of the functioning of a society, causing widespread human, material or environmental losses, that exceeds the local capacity to respond, and calls for external assistance. Disaster often co-occur with public health problems since they (1) can lead to unexpected number of deaths, injuries and illnesses that exceed available health resources, (2) may destroy local health infrastructure and emergency response capabilities, (3) may cause adverse effects to the environment, (4) may cause large populations to move.

In particular in the area of disease surveillance, a large variety of surveillance systems exist (e.g. HealthMap,[5] BioCaster,[6] SurvNet [23]). They monitor different data sources for information on public health threats. Disease surveillance systems aim at supporting the earlier recognition of significant health events. In general, these systems are not specific enough to confirm the presence of any particular disease [45]. Data routinely collected for surveillance systems comprise medication sales, reports of illness from nursing homes, reports of chief complaints from emergency medical departments. These data sources may provide an early indication of a health event, but they do not specifically provide a signal or alert. A concrete disease surveillance system will be introduced in Chap. 5.

Beyond disease surveillance, disaster surveillance systems are in place. The Global Disaster Alert and Coordination System (GDACS[7]) provides alerts about natural disasters around the world (e.g. earthquakes, cyclones, floods, volcanos). Information on such events is picked up from multiple sources and automatically analysed by a computer program to determine the likelihood of humanitarian intervention. News, damage maps and humanitarian situation reports are collected automatically from over thousand online sources. The list of events published by GDACS is compiled from information provided by GDACS partners or from public information syndicated on the Internet. Data from seismic detectors for

---

[5] http://healthmap.org/en/

[6] http://born.nii.ac.jp/

[7] http://www.gdacs.org/

example from the Geoforschungszentrum[8] in Potsdam, Germany, are made public using telecommunication or the Internet.

These examples show that information of various kinds of disasters can be collected by disaster surveillance systems. They target at providing early warnings and supporting the continuous monitoring of disaster events. Disaster surveillance systems mainly target at supporting officials in making decisions on actions to be taken. Nevertheless, there are scenarios, where also normal people could benefit from a disaster surveillance system. For example, when a serious flood happens like the one in Thailand in 2011, a surveillance system could help the population in getting information on the situation in a certain region or information on states of roads, where to go etc.

### 2.3.4 Traffic Surveillance and Management

The objective of traffic surveillance and management is avoiding current traffic jams, improving security of road users and improving the efficiency of the existing traffic infrastructure. Traditionally, systems for traffic surveillance rely upon sensors that collect data about the current traffic situation. These might include sensors that are located in the road surface or data provided by traffic lights, emergency telephones etc. These sensors provide numeric values to a traffic monitoring system.

There is also ongoing activity to exploit additional sources of information, mainly social media tools, for traffic surveillance. BNE Traffic[9] is a research prototype coming from SAP Research. The system identifies tweets tagged with the tag *bne-traffic*, then extracts the geolocation information using textanalysis technology and places the information on the appropriate place in a Google map. In this way, the system monitors the traffic as it is reflected in the tweets.

Social networks as sensors are also exploited in the area of traffic detection and monitoring. Traffic AUS[10] proposes social networks for car drivers. The data produced by drivers about the traffic situation can be exploited by other drivers for real-time navigation [62].

Users of traffic surveillance systems can be the police, road assistance services or any other person involved in road traffic. A police station could get alerts on traffic jams or of accidents, maybe faster then through other channels. Road assistance services can benefit from surveillance systems by getting alerts on cars that need their support. For other persons involved in road traffic a surveillance system could provide information on problems in traffic, on destroyed or occupied roads.

---

[8] http://www.gfz-potsdam.de/geofon/

[9] http://www.appolicious.com/tech/apps/104316-bne-traffic-sap-research

[10] http://itunes.apple.com/au/app/austraffic

# Chapter 3
# Information Sources for Surveillance

The input to surveillance systems is data in which patterns are identified that provide hints to abnormal or aberrant behaviour. Depending on the application domain and surveillance task, surveillance systems collect this input from various information sources. In this chapter, an overview on possible sources of information is given. Collected data can be structured or unstructured. This book focusses on the processing and surveillance based on unstructured data. For completeness reasons, a brief overview on sources for structured data is provided in this section.

The following questions will be addressed in this chapter:

- Which data is available for surveillance purposes?
- What are the characteristics of this data and what kind of information is provided?
- Which challenges need to be considered when exploiting this data for surveillance?
- How timeliness and accurate is the data for the task of surveillance?

The challenges comprise on the one hand challenges for data collection, data stream processing, and data management. On the other hand, challenges result from the nature of the data and makes demands on follow-up processing in surveillance. An overview on these challenges will be given.

Table 3.1 provides examples for structured and unstructured data sources, their characteristics and implications for data processing technologies in surveillance systems as they are described in the following.

## 3.1 Structured Data

Structured data is data stored in a structured way, for example following a specific data model. The data consists normally of numerical values and is thus uniquely interpretable.

K. Denecke, *Event-Driven Surveillance*, SpringerBriefs in Computer Science,
DOI: 10.1007/978-3-642-28135-8_3, © The Author(s) 2012

**Table 3.1** Characteristics and examples for surveillance data

| | Structured data | Unstructured data |
|---|---|---|
| Characteristics | • structured, e.g., numerical values<br>• timeliness, requires sometimes manual data entry<br>• continuous stream<br>• reliable, precise | • unstructured, i.e natural language text<br>• timeliness<br>• update frequency often unknown<br>• not always reliable |
| Implications | • no preprocessing or data preparation required<br>• sophisticated data management necessary<br>• statistical algorithms are directly applicable<br>• data stream management necessary when processing high frequency sensor streams | • data management necessary to cope with huge amounts of data<br>• textual analysis and information extraction required to get structured, processable data<br>• sophisticated filtering necessary<br>• domain knowledge for interpretations and natural language understanding required |
| Examples | GPS position, number of incoming emergency calls | Weblogs, Online news |

## 3.1.1 Sources of Structured Data

A primary source of structured data are physical sensors. They monitor a specific target and provide the sensed data. Sensory devices include infrared, chemical, biological, acoustic, magnetic, or motion sensors [71]. Chemical and biological sensors are used to monitor gas or toxic emissions and possible medical or biological outbreaks. Motion sensors trigger other, more advanced surveillance devices. Infrared sensors are for example used to detect weapons. Thermometers are sensors that provide a numerical value which is the temperature measured. Another example are training computers that collect the heart beat frequency, the GPS position, duration of paces and others.

Each of these sensors provides data, either as a continuous stream or with an update frequency that is previously unknown. For example, the update frequency of a glucose monitor depends on the patient who needs to make the test from time to time. The single values measured by the sensor are of importance for the surveillance system. Normally, all measured values are of relevance.

Structured data that reflects human behaviour can among others be derived from frequent flyer accounts, surfing history, toll-road systems, or credit card usage records. Data collected by commerce and business is another source of—normally structured—information. Commercial data includes for instance numbers of product sales, such as healthcare product sales that are considered as information source within public health surveillance [69]. An increasing number of incoming medical call centre calls or of the number of drug prescriptions can for instance indicate a potential health event [73].

The Web can also be considered as sensor and provides structured and unstructured information for surveillance. Patterns of informal utilisation can be used to detect events. Among others, the following indicators could be useful for monitoring systems that exploit Web data:

- number of hits to a website from some region for some period,
- number of requests to specific websites,
- number of queries to Internet search engines that contain some seed word and thus reflect the situation or object under monitoring.

Studies already showed that the number of flu-related Google queries correlates to the number of influenza cases [29]. This observation is exploited within Google Flu trends,[1] that uses Web search queries for monitoring the influenza season. A similar procedure has been applied to detect hints to Dengue fever outbreaks through analysis of search behaviour.

In addition, Web access logs are used for surveillance purposes. Researchers showed that the number of page accesses to a health related website for articles on influenza and influenza-like illnesses correlate with the reported cases. Thus, this data can be considered for surveillance as well. Due to the number of possibly relevant websites this is still an ambitious organisational task.

Additionally, there are sensors that allow to track human reactions and movements including motions, speech, face-to-face interaction, or location [55]. They provide for example information on whether a person is sitting, standing or walking, about length of speech, speaking speed etc.

In summary, there is a large variety of sensors and data sources available that provide structured data. The data can be exploited for surveillance purposes. However, there are several challenges to be considered when using the data in surveillance systems. They are described in the following.

### 3.1.2 Characteristics and Implications for Surveillance Systems

As the examples before show, structured data is characterised by its structured nature, i.e. numerical values or Boolean values are provided. These values can be directly compared to each other or to some predefined baseline or threshold. The data is collected continuously in fixed time intervals. Apart from inaccuracies during the measurement, the data is in general precise. From these characteristics, some implications for surveillance systems result.

Since the data is normally numeric, statistical methods for surveillance can be directly applied to analyse volumes of data. Depending on the update frequency, the amount of collected data can be very high. Thus, the aggregation process and abstracting of data is necessary as part of data management.

---

[1] http://www.google.org/flutrends/

Single sensors might have limitations or might not be able to monitor the situation or object as a whole. Thus, by integrating results from several sensors complementing information can be received. The data from one sensor can also be used to verify information collected by other sensors. A comparison of data from different sensors is often simple due to the fact that numerical values are provided. The only prerequisite is that the sensors provide the data in the same intervals and are measured in the same measurement unit.

The data collected by a sensor can be directly transferred from the sensor to the surveillance system. Thus, the data is available in real time and changes can be recognised as soon as they occur. For instance, a blood pressure meter could pass the data directly to a surveillance system for analysis or the GPS location of a mobile phone is immediately available. This kind of surveillance data can be used either at or close to the time it was generated. Sometimes, structured data is not collected automatically by a sensor but relies upon manual data entries. In surveillance of infectious diseases for instance, physicians are obliged to report to the health organisation when they diagnose a specific disease. In this case, timeliness of the data is often not given.

Structured data and its interpretation and analysis is considered to be reliable. Clearly, when a sensor is damaged it can provide also incorrect values. But in general, the reliability and trustworthiness of this data is high.

When Web usage data is used for surveillance purposes, privacy issues need to be considered. Web access logs provide among others information about the IP address of a user, time stamp, the action requested by the user (e.g., request for a specific article or for a search), a status (i.e., whether the user experienced a success, failure, server error or others), the transfer volume and the referring URL. Misuse of this data need to be avoided.

In summary, considering structured data in surveillance systems requires sophisticated data management techniques. The analysis as such is easy to realise due to the structured nature of the data. Normally, the data collected is reliable and precise and can therefore be well interpreted and analysed. Nevertheless, for some surveillance tasks, no structured data is available or it could be complemented by unstructured data to provide a more complete picture of the object or situation under monitoring.

## 3.2 Unstructured Data

Unstructured data used for surveillance is mainly free text written in natural language. Of course, documents are often structured into sections, paragraphs and sentences. However, the data, the content as such is not directly machine-processable.

The variety of unstructured data that can be considered for surveillance comprises official reports and documentation (e.g., governmental reports or reports from health organisations), research publications, news articles, advertisements etc. Beyond, textual data provided by enterprises represent another set of data sources (e.g., earnings reports, new product announcements, advertisements). More specifically, we can distinguish:

- Official documents from institutions and enterprises: reports, advertisements, announcements.
- Media: newspapers, magazines, radio, television.
- Web-based communities and user generated content: social-networking sites, video sharing sites, wikis, blogs, forums, twitter, reviews and folksonomies.

In all these media, information about latest news, current topics, opinions and experiences is provided—information that might be of interest for monitoring and surveillance. Depending on the surveillance task, only a small part of available documents is of relevance. Additionally, for early warning surveillance systems, the novelty of content is crucial. A huge source for content that is normally up-to-date is the Web. The next section will describe in more detail the kind of Web content that is available. We will describe to what extent it is useful for surveillance and monitoring systems.

## 3.2.1 The Web as Data Source

Given the development of the Internet in the last years, there exists an emerging emphasis that tries to exploit data from the Internet for surveillance purposes [30]. Monitoring the Web search behaviour, for instance, helps to learn about preferences and current user interests. One example is Google Flu Trends that monitors the influenza season by analysing Web queries for frequent patterns. A similar approach has been tested to detect outbreaks of Dengue fever [11].

Beyond the search behaviour, the content provided in the Web represent a new form of surveillance data. Beyond online news, informal or social media data such as Twitter becomes more crucial for monitoring. This data is interesting for surveillance purposes since it provides first hand information or information that is unavailable elsewhere. Normally, those who experience a change of situation send their impression or observation nowadays via social media tools. Making phone calls is too old fashioned since we can reach much more people by using social media tools. An example of an early information through social media is the death to Osama bin Laden. The first on-the-ground testimony that Osama bin Laden has been killed came from an IT consultant in Abbottabad, Pakistan via Twitter (see Fig. 3.1). He sent the story of the attack on Twitter long before the president's announcement and before any major news stories were published. This shows: latest news and observations are posted through the Web.

More generally, content available in the Web can be broadly placed into two categories. Users can either generate their own content, or alternatively annotate the content created by others. To make the distinction clear, we refer to the former as (*user-generated*) content, and the latter as *metadata*. Further, content can be provided by multiple authors collaboratively (referred to as *collaborative content*) or by one

@ReallyVirtual
Sohaib Athar
Helicopter hovering above Abbottabad at 1AM (is a rare event).
23 hours ago via TweetDeck
Favorite
Retweet
Reply

**Fig. 3.1** The first testimony on the death of Bin Laden were sent via Twitter

**Table 3.2** Examples for categories of Web content

| | Collaborative | Non-collaborative |
|---|---|---|
| Content | Wiki | Weblog, Micro-blog, Video-Blog, Websites, Online news |
| Metadata | Social networks | Review and rating portals, question and answer portals |

single author or author group (referred to as *non-collaborative content*). Table 3.2 provides some examples for these content categories which are described in more detail in the following. Beyond, it will be described to what extent the sources are relevant for surveillance purposes.

### Collaborative Content

The best example for content collaboratively produced by humans in the Web are *Wikis*. They offer the opportunity to get information about specific topics. Wikis can be either domain-specific, such as medical wikis (e.g., Radiopaedia,[2] Flu Wiki[3]). They can also have a broader scope as we all know from Wikipedia. With respect to surveillance, wikis are less important since they are often not reflecting the situational changes in time. This is due to the fact that providing sophisticated descriptions of some topic of interest takes more time. Beyond, the collaborative work requires more time to produce content. The situational changes that are reflected by wikis are valid for a longer period.

### Non-Collaborative Content

Among the non-collaborative Web content there are for example weblogs, or microblogs. *Weblogs* are personal diaries that are made public in the Web. The authors provide news about their personal life and experiences, and discuss other topics of interest, things that happened etc. Professional blogs are also available, where information on a specific domain is provided such as latest news from the medical or financial domain. Health professionals are posting, for example, about health issues of current concern (e.g., vaccination campaigns, disease outbreaks).

[2] http://radiopaedia.org

[3] http://www.fluwikie.com

*Micro-Blogs* like Twitter are another source for distributing, communicating and sharing information among people. The content is limited to 160 characters which leads to a specific way of writing (e.g., lots of abbreviations, emoticons etc.). The information shared can be everything including messages about current feelings, activities, latest news and so on.

Microblogging services are used for surveillance already. For example, Sakaki et al. [63] exploit Twitter users as social sensors for getting hints to earthquake shakes. Since users are tweeting about the earthquake, its occurrence can be detected promptly, the dimensions of damages can be judged, simply by observing the tweets. An important characteristic of microblogging services is their real-time nature. Another characteristic is that sharing has a lifecycle of information production and consumption that is rapid and repetitive [65]. Microblogging is increasingly considered as a means for communicating in emergencies because of its growing ubiquity, the immense speed of communication and its accessibility [72]. For example, when an earthquake occurs, people may tweet about the earthquake before any news agency can write an article about it and even before the official notification has been released. The ability of identifying global trends and events by Twitter is the base of Twitter-based applications such as Tweettronics,[4] oriented to identify market trends and brand awareness for marketing purposes.

Due to its huge number of users (in March 2011, BBC estimated the number of Twitter users as around 200 million worldwide and it is still growing), also a large number of sensors is available, under the assumption that each user can be a sensor. In contrast to blogs or forums that are updated once every several days, Twitter users write tweets several times a day. The large number of updates results in numerous reports related to real-world events [63]. They include presidential campaigns, parties, but also disastrous events such as storm, fire, traffic jam, disease outbreaks and earthquakes. Several studies showed that social media is used by people to gather and disperse relevant, useful information on disaster [37].

In summary, non-collaborative content is an important source of information for surveillance purposes. This informal way of providing information supports early reporting when situational changes happen.

## Collaborative Metadata

Collaborative metadata in the Web comprise mainly social networks. Content communities and *social networking sites* enable people with similar interests to connect. More specifically, people can share information in order to empathise or learn from each other.

For surveillance purposes, messages that are exchanged in such networks can be of interest. Similarly to communication through microblogs, messages are distributed within social networks. These messages can also reflect situational changes. Beyond, the building of larger groups might indicate group forming. Thus, from the user

---

[4] http://www.tweettronics.com

behaviour in such networks conclusions could be drawn for surveillance as well. However, this is still a research topic and has not been studied well so far.

In summary, collaborative metadata can provide information on events and situational changes that are relevant for surveillance.


**Non-Collaborative Metadata**

Single authors or author groups provide metadata among others in forums and review or rating portals where they are commenting, rating and asking questions.

*Forums and Query-Answer portals* (Q&A) offer the opportunity to post queries or engage in discussions. Expert forums enable users to get a qualified answer to a question regarding a specific topic. In Q&A portals, people's objective of posting is mainly to receive information. Depending on the portal, answers can be provided either by professionals or by the general public. In particular forums are an interesting source of information for surveillance since hot topics are discussed there. This content could be used for market analysis and to get product or service feedback.

Similar information is also provided through *reviews and rating portals*. They enable users to present experiences with some product, for instance on drugs, software, hardware etc. It reflects their personal feelings and impressions. In summary, non-collaborative metadata provides a resource of personal feelings and impressions that can be useful for surveillance.

We have seen that various Web sources can provide information for surveillance systems, some more than others. Due to the immediate reporting through social media tools, the first hints to an event are often visible in the social web and distributed through social text streams. Social text streams are collections of informal text communications over the Web that arrives over time. Each piece of stream carries information about real world events. In this sense, social text streams are sensors of the real world [62, 78]. More generally spoken, a social sensor is any source of information that can be identified in Web 2.0. In such sources, situations and facts about users are expressed either by the user himself, by others or just through the interaction with these tools. Examples of social sensors include Twitter posts, Facebook status updates or pictures posted on Flickr.

In this context the term *Open Source Intelligence* has been introduced. It refers to the methods that involve finding, selecting, and acquiring information from publicly available sources and analysing it to produce actionable intelligence. Even what a user searches on some Web search engine or what he buys can be implicitly considered as a sort of social sensor.


## *3.2.2 Characteristics and Implications for Surveillance Systems*

Unstructured data is often natural language text. The nature of this data leads to challenges to be addressed by a surveillance system. They concern collecting and filtering, linguistic processing, and text interpretation. In the following the challenges and requirements will be described in more detail.

## Challenges for Collecting and Filtering of Data

Social sensors, as introduced before, have differing activity; their behaviour is unpredictable. Some of them are very active, others not. A sensor could be inoperable or malfunctioning sometimes (e.g., the user is sleeping and does not send any messages). Consequently, social sensors are not necessarily providing a continuous, same size stream of data as ordinal physical sensors do. Further, social sensors are mostly only active for specific events: People are only reporting about events of their interest, taking place in their environment etc. These characteristics of social sensors need to be considered when interpreting their data.

Given the huge amount of social sensors and—accompanied by this—an immense amount of Web content available, it is important for a surveillance system to collect all relevant content. It is clear that not everything can be collected since processing would take too much time. Thus, the designer of a surveillance system in collaboration with the potential users of the system needs to carefully select the sources that should be monitored. It is crucial to find a right balance between (1) having enough sources under monitoring in order not to miss something relevant and (2) to collect not too much irrelevant data to ensure timely processing and reduce the number of false positives.

Considering microblogging services or other online data such as Twitter or blogs for surveillance purposes is challenging. Social media, in particular microblogs, are short-lived, require quick decisions on how to collect which data while an event is in progress [72]. For example, when the first cases of swine flu occurred, the term "swine flu" was not yet used. Thus, searching for relevant tweets with the term "swine flu" would not have led to the desired success. Further, the terminology changed several times during the progress of the event (e.g., it was called "Mexican flu", "swine flu", "Influenza A"). Decisions about data collection need to be made before any information about a potential event might be available. A surveillance system should therefore be flexible enough to adapt to changing terminology or new surveillance tasks.

## Challenges for Linguistic Processing

The unstructured nature of textual data makes it necessary that it is processed and transformed into a structured form before it can be analysed and interpreted. It requires advanced computational techniques to effectively analyse and exploit the data for surveillance. Processing means that relevant pieces of the text are extracted, and stored or made available in a structured manner.

The various sources of unstructured data make different demands on the processing. Online news, blogs and forum postings are normally written in complete, grammatically correct sentences. However, it might occur that only small parts of the text are relevant. In particular in blogs, authors are changing topics within one posting very often. Therefore, relevant pieces of a text need to be identified beyond selecting whole documents as relevant. In contrast, microblogging services restrict the length of messages, which leads to an intensive use of abbreviations and shortcuts which is difficulty to handle by natural language processing tools. For all sources it holds true that for surveillance purposes the data is current and thus, time-stamped.

The following linguistic characteristics of natural language text need to be addressed within the linguistic processing step:

*Linguistic variation.* Natural language provides the possibility to express the same idea with different words. Derivation and inflection change the form of a word while the underlying meaning remains the same. Inflection changes a word's form, such as pluralization or the tense change of a verb. Synonymy allows us to use other words for expressing the same. In medicine, there are for example several possibilities to express the symptom *shortness of breath*: *dyspnea, short of breath, dyspneic.*

*Polysemy* refers to the phenomenon where terms have the identical linguistic form but different meanings. A frequent type of polysemy are acronyms and abbreviations.

*Negations* can be explicitly indicated by negations terms such as *no, without, denies.* Beyond, implicit negations occur as well as expressions of uncertainty. For example, we can find twitter messages with texts like *I have fever and cough. Maybe I got an influenza.*

*Contextual information.* For surveillance, it is necessary to distinguish between past and present events. Thus, contextual information is crucial to be linked to the data. Further, it helps to create a whole picture of an event by connecting a person or object to a location and additional things relevant for describing an event that occurred. Often, information is not expressed explicitly but implicitly. To understand such implications, i.e. to make inferences, domain knowledge about words, combinations of words etc. is required.

*Coreference.* Natural language offers the opportunity to distribute information on several sentences. The reference is made by using pronouns or expressions similar to the original word (e.g., Mr. Miller, John Miller, John). Again, to get a complete picture, the information needs to be combined.

For some of these challenges software tools and algorithms are already available that support the automatic processing of natural language text (see Sect. 4.3). Most of the existing algorithms have been developed for processing news articles and documents with complete sentences. It is still challenging to apply them to social media data and to achieve good results given the peculiarities of this data source.

## Challenges for Text Interpretation

Interpretation of the meaning of a natural language text requires knowledge of the domain. For automatic interpretation, this knowledge is normally provided to a machine through a terminology or ontology. Thus, when building a surveillance system that processes unstructured data, a knowledge base is necessary for the domain under consideration (see Sect. 4.5).

The interpretation process itself considers the relevant or extracted parts of the documents. Methods for interpretation comprise simply counting the documents or extracted information entities or interpretation rules. In addition, the amount of data available makes it necessary to abstract and aggregate the information within the interpretation process.

An advantage of getting data through social sensors is that it does not require official reporting. Nevertheless, the reliability of the data or the interpretation is sometimes questionable. Consider, for instance, the use of web search behaviour to draw conclusions on the current health status. Who is really searching for information on a disease: The infected person or the one who read the latest news and just wants to get an updated information? The sensitivity of such data is unclear. A high false-positive rate could increase the workload because verification is necessary. For these reasons, it is crucial in surveillance systems that are based on social sensors, to provide the analyst enough information to specify the confidence of the reported information. This includes information about the person or group that published the information or whether there have been other persons reported about the same event.

In summary, surveillance systems that rely upon unstructured data need to deal with an uneven data quality (missing, short, uninformative text), scalability, dynamic data stream of event information, and an unknown number of events and information which is difficult to estimate. This looks certainly extremely challenging. However, the next chapter will show that there are tools and methods that can help.

# Chapter 4
# Surveillance Methods

When designing and developing a surveillance system, we first have to decide on the target of monitoring. Then, data sources suitable to monitor this target need to be identified. Finally, appropriate data processing and analysis methods are combined and integrated in a surveillance system.

In general, a surveillance system works following the flow as presented in Fig. 4.1: data is collected regularly by some data collection methods. The data is then processed by data analysis and aggregation techniques. On the aggregated data, event detection methods are applied which produce signals that are shown through the user interface to the user.

As previously mentioned, the focus of this book is on surveillance systems that process natural language as input. Thus, the focus in this chapter is on methods that are necessary to enable surveillance from unstructured data. Given the nature and requirements of a surveillance process, the event-driven architecture paradigm is a well-suited software design for surveillance systems. This chapter will therefore start with the introduction of the principles of event-driven architectures. Then, an overview on methods and technologies exploited or required in the single processing steps as outlined in Fig. 4.1, is provided.

## 4.1 Event-Driven Architectures

### 4.1.1 Characteristics of Event-Driven Architectures

Surveillance systems aim at obtaining and providing information on threats or changing situations as early as possible. A main requirement is thus the processing of incoming data in real time. Systems following the principle of event-driven architectures are designed to support this kind of processing and allow for monitoring and responding to conditions in an environment in real time.

K. Denecke, *Event-Driven Surveillance*, SpringerBriefs in Computer Science, DOI: 10.1007/978-3-642-28135-8_4, © The Author(s) 2012

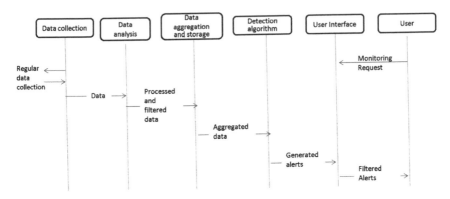

**Fig. 4.1**  Information flow in surveillance systems

Event-driven architectures have been proposed as an architectural paradigm for event-based applications. The basic principle is that the behaviour of the system is orchestrated around the production, detection and consumption of events [9]. In this context, an event is essentially a significant or meaningful change in state. It is an occurrence or happening which is significant for and consumed by a component of a surveillance system. Events are characterised by their type, the time they occur, their number of occurrences, source and other elements [15]. For instance, an event can be some message, token, pattern that can be identified within an ongoing stream of monitored inputs, specific error conditions or signals, thresholds crossed, counts accumulated etc. These are all events that can be detected by a computer and that can reflect real world events such as a disease outbreak or a traffic jam. They must be transformed, classified, aggregated and evaluated to initiate appropriate actions.

A central characteristic of event-driven applications is their composition of three basic technologies that support sensing, analysing and responding (see Fig. 4.2). The *sensing part* requires information about the object under monitoring, i.e., it obtains the data that defines the reality. The sensing part produces *raw sensor events*. These events are too fine-grained and still uncorrelated. Thus, the *analysis part* correlates these events and maps them to domain concepts (e.g. events referring to one disease). Further, it estimates the current state as reflected by the domain events and compares the real state with the expected state of these events. The *responding part* performs some action as response to the event. All three parts can be realised as a combination of single processing components.

Systems that are not event-driven rely upon an active data transmission to the system. In contrast, in an event-driven architecture, data is continuously collected and the arrival of data triggers some analysis [69]. In regular time intervals, the system checks for availability of new data. Data is preprocessed and stored and made available for event detection. The components and services are loosely coupled. A (system) event is transmitted from one service to another (in general, from a provider to a consumer). The consumer's role is to perform some action. It might

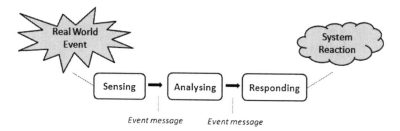

**Fig. 4.2** The principle of event-driven architectures

just have the responsibility to filter, transform and forward the event to another component or it might provide a self contained reaction to such event.

In summary, the main characteristics of event-driven architectures are as follows:

- Event-driven architectures comprise a sensing, analysis and responding part.
- Events are transmitted and communicated as autonomous messages between components.
- Information can be propagated in near real-time throughout a highly distributed environment.
- The software components or services are loosely coupled.

### 4.1.2 Processing Steps and Components

Event-driven surveillance systems work in three steps:

1. **Monitoring and detecting**. Relevant information is detected and collected via sensors or other information sources. The detected information is interpreted as events.
2. **Processing and analysing events**. Events are aggregated, correlated, classified or rejected within the analysis phase. The analysis tries to identify patterns in event streams that refer to certain relations and dependencies between events.
3. **Responding to events**. Detected patterns allow to react timely, including sending alerts, service calls, initiating a reaction from humans or even generating higher order events.

In event-driven architectures, the monitoring is realised by sensors that monitor the environment and generate event messages that contain information about the attributes under monitoring. For instance, a sensor may scan a Web site and then send the information it extracts in an event message. The sensor would be a software module performing the extraction. The Web site itself is not a sensor, but rather an attribute to be monitored.

The event message provided by the sensor is received by a module that performs some action based on the events it receives. Such action could be the sending of

E-Mails to inform users about the detected event. For instance, a responding module could be a software module that gets event information about patients in a hospital and pages doctors, or it could be the pager itself.

In the event processing and analysis step, often event-condition-action rules are exploited which comprise three parts: (1) the event that triggers the rule invocation; (2) the condition that restricts the performance of an action; (3) the action executed as a consequence of the event occurrence. The following example shows pseudocode for an event-condition-action rule.

---

**Pseudocode for an Event-Condition-Action-Rule**

WHEN <Event>
IF <Condition 1>
    THEN <Action 1>
...
IF <Condition n>
    THEN <Action n>

---

More specifically, the analysis part of the surveillance system continuously evaluates the *when* clauses of the when-then rules by fusing information from all the sensors. If a *when* clause is fulfilled, the responder executes the *then* clause.

Accordingly, an event-driven surveillance system consists of several components including:

- a data collection system,
- routines for data aggregation and storage,
- detection algorithms that monitor for abnormal patterns,
- a user interface that displays information.

In Sects. 4.2–4.6 methods that can be exploited within these components will be introduced. Table 4.1 provides an overview on surveillance subtasks and related technology components that supports this subtask.

## 4.1.3 Benefits

Applications and systems following the event-driven principle facilitate more responsiveness, because event-driven systems are—by design—more normalised to unpredictable and asynchronous environments [32]. Event-driven design and development provides the following benefits [32]:

- Allowing to assemble, reassemble and reconfigure new and existing applications and services easily and inexpensively,
- Enabling through component and service reuse a more flexible development environment,

**Table 4.1** Surveillance subtasks and supporting components of a surveillance system

| Surveillance subtask | Supporting technology |
|---|---|
| Monitoring and detecting | |
| | • Data collection |
| | • Data preprocessing (filtering, linguistic analysis etc.) |
| Processing and analysing | |
| | • Event detection algorithm |
| | • Event aggregation |
| | • Event interpretation |
| Responding | |
| | • User interface |

- Allowing easier customisation because the design is more responsive to dynamic processes,
- Allowing to build a more accurate and synchronised system that is closer to real-time changes.

In contrast to systems, where callers must explicitly request information, an event-driven architecture provides a mechanism for systems to respond dynamically as events occur. These characteristics and benefits make the event-driven architectural design paradigm well suited for its application in surveillance systems.

## 4.2 Data Collection Methodologies

In surveillance systems data is collected through physical or social sensors that monitor a situation, person or environment. As the sensing part of a surveillance system, the data collection directly influences the output of the surveillance system. Since most of the unstructured data relevant for surveillance is distributed through the Web nowadays, we will focus on data collection from the Web.

For unstructured data as it is provided through social sensors such as blogs, mircoblogs, forums, web pages the data needs often to be actively collected. This can be realised among others by crawling web pages, or using specific application programming interface (API) to get access to the content of a relevant information providing service. A brief overview on possibilities to collect information provided through the Web is given in Sect. 4.2.1. When considering services such as Twitter that provide data streams and thus, amounts of data are arriving every minute, several challenges need to be addressed. These challenges for data stream management are summarised in Sect. 4.2.2.

### 4.2.1  Web Crawling and Collecting Social Media Data

A *Web crawler* is a system that downloads whole pages from the Web [10]. Such system is mainly used to copy visited Web pages for indexing purposes or other kinds of processing. Web crawling is additionally exploited for creating an index for general Web search, for archiving Web content or to analyse websites.

A Web crawler starts from a set of seed pages that are first downloaded and then searched for additional links. Pages to which the links are referring are again downloaded and so on until some stop criterion is met. There are several open source Web crawlers available, for example NUTCH,[1] WIRE,[2] WebSPHINX,[3] ht://Dig.[4]

Within the context of event-driven surveillance systems, a Web crawler can be considered as Web content monitor. It provides event messages (i.e. the content it collects) that contain information about the attributes it is monitoring.

While a Web crawler focuses on collecting and downloading Web pages, other tools are necessary when considering content from blogs or other social media tools by a surveillance system. The Twitter application programming interface[5] (API) allows to collect Twitter messages. Users can obtain real-time access to tweets in sampled and filtered form. The API is HTTP-based and GET, POST, DELETE requests can be used to access the data. With Spinn3R[6] blog pages can be collected automatically. It indexes the full HTML of weblog posts. Beyond, it extracts the content by excluding sidebar and chrome.

For collecting data with either a Web crawler or one of the other mentioned tools, seed pages or keywords need to be specified. The tools are searching for content matching the provided keywords or start the collection process from the seed pages. For surveillance purposes, such restriction of the content collection methodology has advantages and disadvantages. Given the huge amount of Web content available, it is only possible to process a (relevant) subset of all data. Normally, this is sufficient for surveillance purposes. Consider for example a surveillance system for the financial domain. Texts dealing with medicine or music would not be of interest for such system.

On the other hand, by restricting the content collection to a set of documents or Web pages matching some predefined keywords we risk that some relevant information might be overlooked. It is therefore crucial to carefully select the keywords used for content collection, since this selection directly influences the composition of the data collection and thus also the output of the whole system. When relevant content is missing, nothing can be monitored or an event or a threat might stay undetected. The same holds true for the web page crawling: selecting a good seed page or a set of good seed pages is crucial to get all relevant content.

---

[1] http://nutch.apache.org/

[2] http://www.cwr.cl/projects/WIRE/

[3] http://www.cs.cmu.edu/rcm/websphinx/

[4] http://en.wikipedia.org/wiki/Ht-//dig

[5] https://dev.twitter.com/docs

[6] http://spinn3r.com/

Another possibility to collect regularly content from the Web for monitoring purposes is to receive updates of webpages or blogs directly through structured data services such as RSS feeds. An RSS feed is a Web feed used to publish frequently updated content such as blog entries, news headlines, audio, and video in a standard-ised format. An RSS document contains either the full text or a snippet and meta information on authorship and publication data. This kind of services helps to resist on the burden to run regularly crawling jobs within a surveillance application. The incoming RSS feed is then the event message for the system to be processed.

As mentioned before, also web queries are used for monitoring and surveillance purposes. Google for example exploits the queries to their search engine to offer Google Flu trends[7] or Google Insights for search.[8] Both tools analyse patterns in queries to the search engine. Providers of search engines often store the queries in query logs. Getting access to these query logs is difficult due to privacy issues. Normally, only the provider itself has access to these logs.

## 4.2.2 Data Stream Management

Another issue to be considered when collecting and processing data for surveillance purposes is the immense volume of data available depending on the source and kind of sensor under consideration. Often, a stream of data is arriving continuously over time. Just consider Twitter as a sensor where hundreds of messages are arriving every minute and millions of tweets are coming in every day. Given this situation, it becomes clear that there is a need for data management, collection and analysis techniques that can keep pace with the volume and velocity of data streams.

A **data stream** is a potentially endless flow of data that arrives in an order that cannot be controlled. Data streams are potentially unbounded in size and the data is generated continuously in real time. Thus, the volume of data is very large and the update rates are high and irregular. Traditional database technologies are not able to deal with the peculiarities of streaming data. Therefore, specific data processing techniques are necessary to monitor and analyse these massive volumes of data in real time. Processing techniques need to:

- scale up,
- deal with a high data rate, and
- deal with an previously unknown amount of data.

To address these requirements, the technologies used for data processing need to be time and space efficient. A possible solution within data stream management is to process only parts of a data stream at one time. Beyond, another challenge is the irrelevance of large parts of the incoming data. On the one hand, this makes new requirements for efficiency and scalability of follow-up processing and storing. On

---

[7] http://www.google.org/flutrends/

[8] http://www.google.com/insights/search

the other hand, the incoming data needs to be carefully filtered to avoid a negative impact on the quality of the follow up processing. Filtering can be realised by text classification or clustering methods (see Sect. 4.4).

Considering the peculiarities of data streams is crucial in all processing steps. It is mainly an issue to scale up existing technologies such as machine learning. For scaling up machine learning algorithms, there are two general strategies. The *wrapper approach* tries to collect data in a batch to which traditional methods are applied. The *adaptation approach* looks for new methods that are adapted to the data stream setting [7].

A subset of data stream management methods is referred to as *data-based techniques* [28]. These methods summarise a data set or choose a subset of it to be analysed. The idea is to reduce the data set to some meaningful subset. Different techniques are available for this summarisation or data set size reduction: *sampling* makes a probabilistic choice of a data item. *Load shedding* drops a sequence of data streams. *Aggregation* computes statistical measures that summarise a stream. For surveillance systems, such data-based techniques might not be the right choice for data stream processing. By aggregating or selecting items from a data stream, a selection of items is made that can lead to the loss of important data. Within surveillance, it is crucial to become aware of even small changes in a data set. These might be overlooked when applying sampling or aggregation.

Another set of data stream management methods is referred to as *task-based techniques* [28]. With these techniques the computational challenges of data stream processing are addressed. Among others approximation algorithms or sliding window algorithms have been introduced. Sliding window algorithms for example, perform a detailed analysis of data for the most recent data items. Only a summary of the old data items is considered for processing. These methods are better suited for surveillance systems. The current or most recent data is of large importance for monitoring and requires deep analysis. We will see later statistical methods used in surveillance that make use of this principle of dealing with data streams.

A more detailed review on data stream technologies is given by Gaber et al. [28].

## 4.3 Data Preprocessing Methods

From the collected data, it is crucial to filter out irrelevant texts, to identify relevant information and to make unstructured text ready for data analysis. We refer to this process by the term *data preprocessing* and describe methods for these three subtasks in the following.

### 4.3.1 Filtering Methodologies

The amount of data collected by sensors can be immense. Thus, it is crucial to prefilter the continuous stream of incoming textual data. One objective of filtering is to reduce processing time at some later stage, but also to avoid information overload. This kind of relevance assessment can be realised by:

- selecting texts that match certain criteria,
- classifying texts automatically, or
- combining both approaches.

A straightforward approach to text filtering is to exploit certain predefined criteria. These criteria can be user-provided patterns or keywords. Texts that match these patterns or keywords are included into the text collection for further processing. Simple pattern matching approaches can be exploited for this purpose. The selection criteria can also be learnt automatically from training material. Machine learning algorithms such as Naive Bayes are applied for this purpose. The training material can be either generated manually by humans or (semi-)automatically. Transfer learning or inductive learning is a research problem in machine learning that focuses on storing knowledge gained while solving one problem and applying it to a different but related problem [76]. Transfer learning can be exploited to label training data automatically based on some additional data source.

For classifying texts automatically, supervised learning approaches or text categorisation approaches are used. They comprise three main processing phases: (1) feature extraction and selection, (2) document representation and (3) induction. Within the *feature extraction and selection* phase a feature set is determined that is used to represent training documents. In the simplest case, all words of a document collection form the lexicon, and in this way, the feature set. To reduce the lexicon and vector size to meaningful pieces that represent a text, feature extraction often applies linguistic analysis to each training document to produce a list of nouns or noun phrases referred to as features (see Sect. 4.3.2). Stop words, i.e. non-semantic bearing words, are excluded from the feature set.

The goal of feature selection is to condense the size of the feature set, and to improve learning efficiency and learning effectiveness [21]. Several methods for feature selection have been proposed including information gain, mutual information, or correlation coefficient. The top $k$ features with the highest feature selection metric score are chosen for document representation.

In the *document representation phase*, each training and test document is represented using the features in the dictionary built in the previous step. For example, documents can be represented as vectors along the dictionary. By representing documents in a similar way, they become comparable.

In the *induction phase*, finally patterns are discovered in the training data that distinguish documents from one category from another based on the set of preclassified training documents. Several learning algorithms are available. Each of them makes assumptions about the nature of the patterns to be found in the data. Thus, an algorithm need to be selected that is well suited for the particular data set. For textual data, and in particular, for processing data streams well-suited learning algorithms include Multinominal Naive Bayes, support vector machines, entropy maximisation and neural networks [75]. The discovered patterns are then exploited to classify unknown data from the test data set.

For text filtering and classification, it is crucial that the underlying keywords and patterns are of good quality, meaning that they allow to distinguish documents. The quality of the training set influences the classification accuracy. Further, the data representation influences the quality of the classifier.

## 4.3.2 Text Analysis and Natural Language Processing

As defined before, event detection is looking for anomalous patterns. Such pattern could be the occurrence of some instances or variables over a given time period. In order to detect patterns and to enable event detection algorithms to make use of the unstructured data in textual documents, the relevant information need to be extracted and converted into a structured format. This is the objective of natural language processing (NLP). It annotates textual data streams with data that reveal types of entities, as well as temporal and geospatial information.

In more detail, NLP refers to automated methods for converting free text data into computer-understandable form. In this section, an overview on NLP tasks and their relevance for surveillance is given. For details on the technologies, we refer to literature on text mining and natural language processing, e.g., [39, 47, 49].

**NLP Tasks**

Natural language processing comprises several tasks:

- sentence splitting,
- tokenization,
- part of speech tagging,
- named entity recognition,
- morphological analysis or stemming,
- linguistic parsing,
- negation detection,
- coreference resolution,
- relationship extraction.

These tasks are described below and in Table 4.2. They can be realised by statistical or symbolic techniques. Statistical techniques use information from the frequency distribution of words within a text to classify and extract information. Classification and clustering algorithms fall into this category (see Sect. 4.4 for more details). Symbolic techniques exploit the structure of the language (syntax) and the semantics to interpret a text. Further, there are hybrid techniques or applications that use both, symbolic and statistical techniques. A detailed overview on NLP techniques is provided by Feldman and Sanger [24].

Statistical NLP techniques have been used among others to address the problem of polysemy. Given a word or phrase with multiple meanings, the statistical distribution of the neighbouring words in the document helps in disambiguating the correct meaning. Beyond text classification, statistical techniques can be used for complex NLP tasks (e.g., extraction of information). Also sentiment analysis (see Sect. 4.3.3) is often realised by means of text classification.

**Tokenization**. Before the actual linguistic analysis takes place, a document is segmented into linguistic units, i.e. into paragraphs, sentences, words, phrases. Tokenizers split sentences into words. This segmentation is a prerequisite for all other NLP tasks. To build a feature set for surveillance algorithms from unstructured documents, at least tokenization needs to take place. Tokenizers often rely upon symbolic techniques (e.g. regular expressions) that use heuristics determined based on statistical techniques.

**Part of speech tagging**. After document segmentation, the text is normally enriched with grammatical information. Each token is assigned to a part of speech (e.g. verb, noun, preposition) which is realised by automated part of speech taggers (POS Tagger). They use either rules or probability distributions learnt from hand-tagged training sets to assign parts of speech. For general English texts like newspaper articles, taggers perform with an accuracy of 96–97%. Unfortunately, when applying these taggers to domain-specific texts, their performance decreases. Training on domain-specific material is crucial to ensure a good quality. With respect to surveillance, the assigned parts of speeches can be used to filter out features when building a feature set. For example, a feature set can be restricted to content-bearing words like nouns and verbs.

**Named entity recognition**. Another NLP task is named entity recognition (NER) that targets at finding instances of some entities in a text. For example person names, locations or organisation names are types for named entities. The recognition can be realised by a lexicon look up or by pattern matching. The named entity recognition process results in instances extracted from a text which are assigned to a predefined set of named entity types. Monitoring can be performed on certain entity types. In a disease surveillance system, for instance, entities of the entity types *disease* or *symptoms* are monitored.

**Linguistic parsing**. After tagging, parsers determine larger units among the tokens. Automated parsers exploit a grammar consisting of rules or probability distributions for breaking down a sentence into its subparts (phrases). A full parser determines the complete hierarchical structure of the constituents of a sentence. A shallow parser only determines chunks such as noun phrases or prepositional phrases. The syntactic structure of a sentence can provide information about the semantic relationship between words. It can become also a feature for event detection algorithms since tokens that are related syntactically can be interpreted as patterns relevant for event detection. Parsing is also a prerequisite for relation detection (see below).

**Negation detection**. This is another important NLP task. Negation is the process by which a negating word (such as "not") inverts the evaluative value of an affective word ( for example, "not good" is similar to saying "bad"). This can be resolved in natural language processing by identifying negating words, and then inverting the value of any positive or negative word within n-words of the negating word, where n is the window of potential negation. The most successful approaches to negation detection exploit symbolic techniques that try to match negation patterns. For surveillance it is extremely crucial to determine whether a fact is negated or not. Just imagine messages such as "The flood is gone." or "No flood anymore". A surveillance system should not generate alarms for these cases.

**Coreference resolution**. Another linguistic phenomenon relevant in automatic text processing are coreferences. In linguistics, coreference occurs when multiple expressions in a sentence or document refer to the same thing. Coreference resolution aims at identifying such references. For example in a text "Mr. Miller" can be referred to by his full name "Jim Miller", by a pronoun "He" or his profession "the doctor". This problem can be considered on document-level and becomes even more difficult, when considering coreference resolution on a cross-document level.

In general, algorithms for resolving coreferences commonly look for the nearest preceding individual that is compatible with the referring expression. For example, "he" might attach to a preceding expression such as "Mr. Miller" or "Jim", but not to "Jennifer". Thus, at the beginning of the process, mentions of entities need to be identified in a text and relevant information about the mentions (e.g. number and gender) are determined. Then, the coreference resolution takes place, realised by different methods including string matching [44], classification [16] or clustering [54]. A current approach in this field is a combination of several approaches that are performed one after another [44]. For example, at the beginning, exact string matches are searched. Second, a more relaxed string match could take place and at the end pronominal coreference resolution is performed [44]. Even though researcher worked on these problems since several years, their F1-measures still do not exceed 60–70% in evaluations. Beyond, most of the work on coreference resolution focused on news corpora. It is still unclear how the approaches behave in other real-world domains or even on social media data.

For surveillance, the cross-document coreference resolution is extremely important since it could help to identify texts that refer to the same event (e.g. a set of tweets that are referring to the same event). Coreference resolution within a document is also crucial to identify all relevant pieces of information that are provided for an event.

**Relation extraction**. Beyond identifying coreference relations, other types of relations between entities exist. More generally, relations are facts involving a pair of individual entities (e.g., *Moby Dick was written by Hermann Melville* with the relation *author-of* between an entity of type *book* and *author*). Thus, relations can be described as attributes of one of the entities. Other examples of relations are person-affiliation and organisation-location. Relations capture much of the connection between entities, and can be used to build entity networks.

Within the task of relation extraction, relations between entities are identified. Most relation extraction systems focus on extraction of binary relations. One possible approach to address the relationship extraction task is to consider it as binary classification problem and apply supervised machine learning. For this purpose, positive and negative instances of a relation need to be available. Then, instance attributes need to be defined in terms of annotations and their features. Based on the feature set, a model can be trained. Features include syntactic features, e.g. the entities themselves, the types of entities, word sequence between the entities. As semantic cue, the path between two entities in a dependency parse tree has been successfully exploited [40].

Another set of approaches to relation extraction are semi-supervised and boot-strapping approaches. Such systems only require a small set of tagged instances as starting point or a few manually created extraction patterns per relation to launch the training process [1, 22].

Relation extraction is of importance for surveillance in two ways. First, the monitoring task can involve relations. For example, the task could be the monitoring of malaria infections in Africa. Then, only texts are of interests where these two entities are mentioned in a certain relationship. Beyond, the detected relations can be used to describe events: extracted relations can be presented to a user to provide a first overview on the content.

**Information Extraction**

A surveillance system often provides information on the event detected in terms of event properties. These properties include entities, their relationships and attributes of entities. They can be extracted from texts by means of information extraction technologies.

Information extraction refers to the automatic extraction of relevant data from unstructured documents. It transforms the data into structured representations, often referred to as templates [14]. Early extraction tasks were concentrated around the identification of named entities, like people and company names and relationships among them. We previously introduced this as named entity and relationship extraction. Later, the tasks became more complex and involved the filling of event templates (Fig. 4.3). An information extraction task comprises nowadays several subtasks, including named entity recognition, coreference resolution etc.

Methods used for information extraction can be categorised along two dimensions: (1) hand-coded or learning-based and (2) rule-based or statistical. A *hand-coded system* requires human (domain) experts to define rules or regular expressions for performing the extraction. In contrast, *learning-based systems* require manually labelled examples to train machine learning models for the extraction task.

On the other hand, we distinguish rule-based and statistical approaches to information extraction. *Rule-based extraction methods* are driven by fix extraction rules, whereas *statistical methods* make decisions based on probabilities. They focus on discovering statistical relations. The latter are more robust to noise in unstructured, social media data, and therefore, more appropriate for surveillance systems dealing with such noisy data. A disadvantage is that a large amount of data is required in order to get statistically significant results.

The rule-based methods base upon a set of extraction rules that can combine syntactic and semantic constraints to extract from a document information relevant to a particular task. For example, the extraction system WHISK uses extraction rules

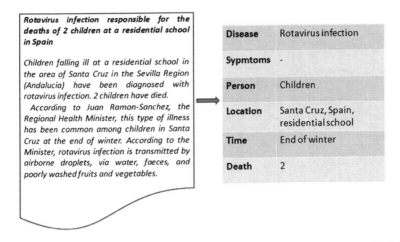

**Fig. 4.3** Information extraction extracts relevant data from unstructured text (*left*) and fills it into a predefined template (*right*)

based on regular expressions [64]. Patterns and fix extraction rules are useful when very specific information need to be extracted [35]. A detailed survey on information extraction methods is given by Chang et al. [12].

## Text Analysis Tools

For the various NLP tasks described before, tools are often already available. Table 4.2 lists some examples. They provide rich linguistic annotations. For example Open Calais[9] is such tool that exploits NLP and machine learning to extract entities from texts and combines them further into facts.

Most existing NLP tools are available for general English and have been tested on news articles. When considering Web data, in particular social media data, additional challenges need to be considered. They include writing errors, common language and slang, amounts of irrelevant content. Further, NLP tools are often domain dependent and require background knowledge from the domain under consideration. The informational content and structure of a domain forms a specialised language called sublanguage [27]. The linguistic characteristics of a sublanguage influence an NLP system's quality and thus need to be considered. For some domains such specialised tools are already available. There are, for example, many NLP tools available for processing medical and biomedical texts. However, for most of the domains specific tools are still missing and non-specialised NLP tools perform badly on texts with domain-specific language.

---

[9] http://www.opencalais.com/

**Table 4.2** NLP tasks and tools

| Task | Description | Tools |
|---|---|---|
| Tokenization | Split documents into token | GATE[a] |
| POS tagging | Assigning a word class to each token | LingPipe[b], Stanford NLP, GATE |
| Parsing | Identify larger units in a sentence | GATE[a], Stanford NLP[c] |
| Named entity recognition | Identify entities of a specific type | OpenCalais |
| Coreference resolution | Identifying references to the same entity | BART[d], Stanford NLP |
| Relation extraction | Identify relations between entities | T-Rex[e], KnowItAll[f] |

[a]http://gate.ac.uk/
[b]http://alias-i.com/lingpipe/
[c]http://nlp.stanford.edu/software/index.shtml
[d]http://www.bart-coref.org/
[e]http://sourceforge.net/projects/t-rex/
[f]http://www.cs.washington.edu/research/knowitall/

### *4.3.3 Sentiment Analysis*

Companies, politicians and other stakeholders become aware that the tone of a body of text, in particular of social media, can influence their business. When for example many people express negative opinions towards a politician in some online forum, it might indicate a need for a future political change. Pessimistic language picked up by market participants might affect share prices. Thus, monitoring the sentiments expressed towards some topic becomes of interest. Sentiments can be either emotions or judgements or ideas prompted or coloured by emotions [8]. The challenge in sentiment monitoring is that sentiments or opinions are often not stated in a clear way in a text. Instead, they are expressed in subtle and complex ways.

Sentiment analysis can be considered at different levels (word-, sentence-, document-level), as two or three-class classification problem (positive, negative neutral) or even with a more fine-grained sentiment scale. To assess the sentiment and opinion expressed in texts, sentiment analysis techniques are available. There are two main techniques for sentiment classification: symbolic techniques and machine learning techniques [57]. Symbolic approaches use handcrafted rules and lexicons. A simple symbolic technique would be to look up words in a sentiment dictionary, count the number of positive and negative words and decide on the polarity of the text based on the frequency of positive and negative words. Various lexical resources are available for this purpose providing words assigned to a sentiment category (e.g. positive, negative) or a score reflecting the polarity of a word. Some of the resources available are listed in Table 4.3.

Machine learning approaches to sentiment analysis rely upon unsupervised, weakly supervised or fully supervised learning to build a classification model from training corpora. As for all classification problems, it is crucial to decide for a well-suited feature set. Several features are commonly used like n-grams or part-of-speech

**Table 4.3** Lexical resources for sentiment analysis

| Name | Description | Language |
|------|-------------|----------|
| General Inquirer[a] | Data is comprised of several hand annotated dictionaries which attempt to place specific words into various descriptive categories. Categories include "positive", "negative" | English |
| SentiWordNet[b] | SentiWordNet assigns to each synset of WordNet three sentiment scores: positivity, negativity, objectivity | English |
| WordNet Affect [70] | WordNet Affect is a linguistic resource for a lexical representation of affective knowledge | English |
| SentiWS[c] | SentiWS is a lexical resource for sentiment analysis and opinion mining | German |

[a]http://www.wjh.harvard.edu/~inquirer/
[b]http://sentiwordnet.isti.cnr.it/ [10]
[c]http://wortschatz.informatik.uni-leipzig.de/download/

or opinion words [57]. Opinion words are words that people use to express a positive or negative opinion. To identify opinion words again the lexical resources like those listed in the table are used.

To train a classifier for sentiment classification, standard supervised learning techniques have been used, for instance support vector machines or naive bayes. A supervised classifier requires a labelled training corpus to learn a classification function. Several corpora for training machine learning classifier have been released [57]. Normally, sentiment classifiers perform with an accuracy between 85 and 90%.

Meanings of words, and in this way the opinion expressed often rely upon the domain and context in which a term is used. Since labelled corpora are not always available for all domains, unsupervised methods can be exploited to label a corpus. In addition, there are weakly supervised methods applied in this field. For example, from a set of seed words with known polarity, an expanded collection of words is determined. Open challenges are still to determine sentiments towards a certain topic, cross-domain and cross-language classification as well as dealing with low quality texts. An introduction into the facets of sentiment analysis, and technologies is provided by Pang and Lee [57].

For surveillance, sentiment analysis is crucial when the surveillance task deals with the monitoring of sentiments for a brand, product or topic. When for example the opinion expressed towards some topic suddenly reverses, this could be interpreted as an event. Beyond, sentiment analysis can be applied to predict political affiliations [51].

## 4.3.4 Role of NLP for Surveillance Systems

In event detection and surveillance systems, the structured information produced by NLP tools is exploited for analysis and interpretation purposes. Detection algorithms require the structured input that is produced by NLP tools. Most frequently used by current approaches are extracted named entities. The frequency of instances of a specific named entity type can for example be used as input for an event detection algorithm (e.g. named entities of type *location*, *disease*, see Sect. 4.4). Another possibility is to exploit the number of sentences relevant for a concrete surveillance task as input for detection algorithms (e.g. the number of relevant sentences). An event classifier could exploit linguistic structures like the parse tree or part of speeches, to identify relevant pieces or sentences of a text. More details on how to exploit the results from NLP tools by event detection methods are described in Sect. 4.4.

## 4.4 Data Analysis and Interpretation

When doing surveillance manually, the following procedures are applied: (a) calculating rates, ratios and proportions to assess the significance of the collected data, (b) comparing these rates with expected values or reference rates, (c) preparing tables, graphs and charts for visualisation. Due to the immense amount of data available nowadays for surveillance and monitoring, the analysis of data can only be realised by means of automatic or at least semi-automatic technologies.

The data analysis starts once data is acquired by the data collection and preprocessing methods presented before (see Fig. 4.4). From the structured data determined by natural language processing and information extraction methods, a feature set can be established. Then, a variety of analytical processes can be applied to this feature set to determine patterns, identify trends or generate signals. In this section, an overview on corresponding methods for event detection, data mining and signal generation is given.

### 4.4.1 Event Detection Methods

An important step within the data analysis is detecting events. Even though there are many similarities, surveillance and event detection is not the same: event detection targets at finding any event when it is reported the first time. This is—of course—also crucial for surveillance systems. Thus, event detection methods are an important part of such systems. But, surveillance systems perform some additional interpretation based on the detected event. It is not sufficient for an event to be new, but also to be relevant and of importance for a surveillance task. Thus, we consider event detection as important subtask of surveillance.

Event detection techniques identify interesting events from streams of texts. An event is interesting for surveillance, when there is a substantial difference between expectation and reality. This can be for instance manifested by an exceeded threshold (see Sect. 4.4.4). Depending on the surveillance task, interesting events can occur rarely, but detecting such an event can have a significant impact within surveillance.

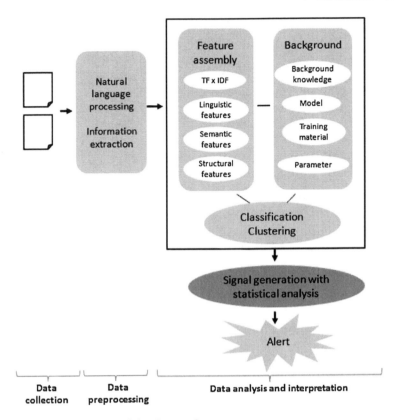

**Fig. 4.4** The event detection and signal generation process

For example, an outbreak of Malaria in Germany will not happen very often if at all. However, even updated information on an event is often subject of monitoring activity. Given, for instance, a serious flood as happened in Thailand in autumn 2011, it is of monitoring interest to identify updates that are related to this mayor event, e.g. when water is coming into other areas or when there is suddenly an increasing number of sick persons in some of the areas.

Practically, an event is referred to something happening in a certain place at a certain time [4, 77]. Known events can be distinguished from unknown or new events. An unknown event is an event that has not been reported before. Imagine for example that an earthquake happens. People start twittering about this new event. A known event has been reported before already. After the earthquake happened, people might provide updated reports on this (now known) event. Known events are considered within event tracking, while event detection technologies normally target at finding the first mentions or hints to events (i.e. new events).

Various approaches to event detection are available. The methods have in common that the task of identifying events is considered as detection of anomalous activity within data streams. Traditionally, event detection techniques base upon a *feature co-occurrence approach* [74]. Another approach is the *event-property based approach*. These approaches and an overview on concrete data mining methods will be presented in the following.

## Feature-Based Approaches

The feature-based approach bases upon co-occurrences of features. For detecting an unseen event in some text, the similarity of features between the new text and past texts is compared. The simplest way of feature-based event detection is to monitor the frequency of words or to detect bursts in word mentions or document appearances, in particular of those words or documents that indicate the occurrence of an event. The high-level idea is to analyse a stream of documents and find features whose behaviour is 'bursty', i.e. the features occur with high intensity over a limited period of time [42]. From times without event occurrence, the normal distribution or frequency of words or pairs of words can be determined. When at some point a word or set of words occur much more often, this might indicate an event. Combinations of lexical similarity and temporal proximity have been chosen as criterion for feature-based event detection [77].

## Event Property-Based Approaches

Improvements of the classical feature-based approach led to event property-based approaches. They rely upon the extraction of event properties that are exploited to analyse the nature of an event. These approaches use a combination of text categorisation or clustering and information extraction techniques [74]. A prerequisite for event property-based approaches is that known events have been annotated with event properties (i.e. information on what happened when, where and to whom). From new texts event properties are extracted automatically by means of information extraction techniques. The extracted properties are compared to the properties from the known events and conclusions can be made whether it is a new event or an update of a known event [61, 74]. A benefit of event property-based approaches compared to feature-based approaches is that event information on location, time, kind of event etc. is directly available.

## Data Mining Methods

Feature-based and event property-based approaches mainly differ in the underlying attributes that are used for analysis purposes. As previously described, these features can be among others term frequencies, lexical features or event properties. The analysis as such is realised by data mining techniques that are applied to feature vectors. Data mining targets at finding correlations or patterns among attributes in large data sets. Patterns of interest are identified using statistical techniques such as regression analysis and Bayesian analysis [75].

Once documents are represented by features determined by the methods described in Sect. 4.3, the data mining task can comprise:

- Association discovery. Similar occurrences within the data are identified and expressed as rules with a confidence factor.
- Classification. Data is grouped into predefined categories based on certain, defined characteristics.
- Clustering. Data is grouped based on automatically discovered similarities.
- Sequential discovery. Repeat occurrences are discovered over a predefined time window.

For event detection within surveillance systems, clustering and classification is applied. The objective of both tasks is the grouping of texts. In the field of data mining, classification is realised with supervised learning approaches, while unsupervised learning approaches are exploited for the clustering task.

**Supervised learning approaches** or classification methods aim at learning a classification model that can predict the class of unlabelled examples. The model is learnt based on examples whose class is supplied. Within event detection, several classification tasks are possible. For instance, one task could be to classify texts or sentences into event types such as "Die", "Attack" or "Meet" [52]. Another task within the new event detection is the classification of texts into the categories "new event" and "old event" [43].

In order to apply classification algorithms, a feature representation of natural language documents has to be produced. The objective of feature representation is to represent a text in a way so that texts in the same category have similar representative values, while texts in every other different category have very different representative values. Features used for classification are terms, noun chunks, lexical information, structural features such as sentence length and position [52]. Beyond, part of speech, terms of specific word classes like nouns or verbs, their morphological stem, the syntactic structure, extracted entities etc. have all been used to classify events [39]. For new event detection with classification, other features have been introduced as well, including a named entity overlap, non-named entity overlap [43]. In event property-based approaches, extracted event properties are used as features.

Once a feature representation is produced for documents under consideration, a classification algorithm can be applied to learn a model. In text classification, support vector machines have proven to be most successful. Based on labelled examples, i.e. texts that have been assigned to one of the possible categories, a classification model is learnt, which can then be applied to unknown texts. Usually, it is impossible to classify a text into a single category. Hence, most classification models provide a membership probability for a text and each category.

When processing streams of data, as it is quite common in surveillance systems, a classification algorithm must meet several requirements [5]:

1. work in a limited amount of time,

2. use a limited amount of memory, and
3. process an element from a data stream as it arrives.

These issues related to data stream management, and possible solutions to address them, have been described already in Sect. 4.2.2.

As a result of classification, we know to which category a document belongs, e.g., whether a document reports about a new event or an old event, or whether a sentence provide information on a certain event type. This information often needs to be further interpreted within surveillance systems. It is not yet sufficient to know that a sentence provides information about an event. Instead, it need to be decided automatically, whether this information is of relevance for the surveillance task. This interpretation can be realised by statistical methods as they will be introduced in Sect. 4.4.4.

**Unsupervised learning approaches** group documents into clusters. State-of-the-art clustering approaches to event detection compute the similarity between documents using the vector space model [61]. Then, a similarity measurement such as cosine similarity is used to assign similar documents in the same cluster [43, 77].

There are two event detection approaches that are realised using unsupervised learning approaches, i.e. based on the document clustering approach [77]: the *retrospective event detection* and the *online event detection*. *Retrospective event detection* approaches target at discovering previously unidentified events in a chronologically ordered set of documents. This detection approach lacks information on novel events. Instead, it exploits unlabelled historical data as learning data and identifies events in this data collection.

*Online or new event detection* tries to identify new events from live data in real time. In more detail, new event detection targets at identifying those documents discussing events, that have not already been reported before. The task is thus to find the documents that mention an event the first time [4]. For retrospective and online event detection the interpretation of the result of the clustering is the same: the clusters are interpreted as events. The documents assigned to a cluster are incidents or indicators for the event.

Besides the feature vectors required as input, clustering algorithms often allow to specify parameters such as the number of events (or clusters) to be detected. Making such decision is difficult in practise since it is unknown how many events might have been reported at some point in time [25, 79]. For this reason, researchers suggested methods to learn such parameters automatically. For example, a probabilistic model for retrospective event detection has been used together with an expectation maximisation (EM) algorithm to maximise the log-likelihood of the distributions and learn the model parameters [79]. In online event detection, a single pass incremental clustering is often applied [4]. During a training phase, a threshold parameter is tuned empirically. For a newly arriving document, the similarity between the document and known events is computed. If the maximum similarity is more than a predefined threshold, the document will be assigned to the corresponding event. Otherwise it will be considered a new event. Since the clusters constantly evolve over time, the features for each cluster need to be updated periodically and features of newly formed clusters need to be computed [34].

For surveillance purposes, alerts or signals need to be generated for relevant events. Therefore, interpretation of the detected events (or clusters) is necessary. A key assumption in anomaly detection is that normal data records belong to large and dense clusters, while anomalies do not belong to any of the clusters or form very small clusters. Thus, an event can be considered of interest if (1) documents do not fit into any cluster, (2) small clusters were created, or (3) low density clusters have been produced.

To visualise clusters for the purpose of user assessment, the cluster centroid is often chosen, i.e. the $k$ most significant terms of a cluster. When the underlying approach is event property-based, these properties can be exploited to represent the detected event. Additional visualisation techniques are presented in Sect. 4.6.

Classification and clustering methods can also be combined in hybrid approaches to event detection: first, a clustering algorithm groups documents. The classifier then predicts which cluster correspond to events [34].

In summary, classification and clustering approaches to event detection have certain characteristics that need to be considered when deciding for an approach in a surveillance system. Several things need to be considered:

- text quality (high quality text (news, reports) or noisy social media),
- background knowledge (training material for classifier, ontologies, number of clusters),
- detection mode (online vs. retrospective).

Another crucial thing to consider are the features used to represent the documents, since this influences the quality and the output of those algorithms.

## 4.4.2 Event Tracking

The event tracking task has been addressed intensively between 1998 and 2004 within the Topic Detection and Tracking initiative (TDT[10]). The main focus of the initiative was on the processing of all broadcast news in multiple languages. Within this initiative, a topic is considered an event or activity, along with all directly related events and activities. A story is a segment of news that includes two or more independent clauses about a single event. TDT focused on five tasks:

- Story segmentation. Segment a source stream into its constituent stories (for radio and TV).
- Topic tracking. Find all stories that discuss a specific topic in multiple source streams.
- Topic detection. Cluster texts that discuss the same topic.
- First story detection. Detect the first story that discusses a topic.

---

[10] http://projects.ldc.upenn.edu/TDT/

- Link detection. Detect whether a pair of stories discuss the same topic.

Event tracking starts from a few sample texts talking about an event (or a topic) and identifies all subsequent documents that discuss the same event [4].

Event tracking can be considered as information retrieval task and can thus be realised with information retrieval technology. The event to be tracked is the query and we are interested in identifying documents that match this query. The event and the documents are represented by features (named entities, term frequencies etc.). By comparing the representations of the query and the documents using a similarity measure such as cosine similarity, documents matching the query can be identified [3, 4].

Event tracking is of relevance in surveillance systems that track the development of events over time. For example when a user wants to assess the risk associated with an event, it might become crucial to know whether the event is still ongoing. Additionally, the first documents reporting about an event might provide limited event information. The additional documents determined within an event tracking process can provide updated or extended event information.

### 4.4.3 Aggregation of Events

To increase the coverage of a surveillance system and in order not to miss any relevant hint to an event, data collected by multiple sensors need to be integrated to get a complete view. This kind of information fusion is crucial to improve diagnosis, decision making and control action. For this purpose, events detected by various sensors need to be integrated or aggregated and relations between events need to be identified. *Complex event processing* methods aim at identifying event patterns which signify correlations between events [46]. These methods identify the most meaningful events within an event cloud, analyse their impact, and take subsequent action in real time (e.g. sending alerts to a user). Relevant knowledge is derived in form of complex events, i.e. in situations that can only be described by a combination of events.

To generate a complex event or aggregated event, simple events need to occur that are related by two factors:

1. Time. There is a timely order between two events (e.g., event A comes before event B),
2. Causality. Two events are related to each other (e.g., event B depends on event A).

If two or more events are related, then they can be aggregated to a complex event where the aggregated event refers to the sum of the single events. Clustering approaches can be used to identify patterns between events. The grouping is performed based on the event properties. Another possibility is to use rules for event aggregation like *If $E_1$ and $E_2$ Then $E_3$* (where $E_1$, $E_2$, $E_3$ are events). The field of event aggregation is still open for research. In particular, applying complex event processing to event data collected from different sources has not yet been considered.

Besides aggregating events detected by different sensors, they can be used to analyse or confirm whether a trend or pattern observed in one source is confirmed by the other data sources. So far, this task of automatic event verification has not been addressed.

### 4.4.4 Interpretation of Detected Events

Detecting patterns that indicate to events is not always sufficient for surveillance systems. In particular, when a system is expected to provide alarms on real world events to a user, an interpretation of the detected patterns is required. It needs to be decided whether the pattern points to something unusual or new. Based on such interpretation of patterns and data streams, a system can decide whether an alarm should be generated.

Interpreting patterns often corresponds to detecting changes in data streams. A change can be gradual or more abrupt. To address this task, various strategies or algorithms are available. Structured data is commonly represented as a sequence of observations recorded at regular time intervals, i.e. as univariate time series. We can apply several methods to such time series to determine anomalous patterns and to generate signals [69]. One simple way to detect anomalies in a univariate time series is to look for an obvious spike. Further, algorithms from statistical quality control [31] can be applied including:

- control charts,
- moving average, or
- CUSUM.

In general, these methods model the background activity, i.e. the behaviour or conditions when no event takes place. This background activity is estimated according to various assumptions such as requiring a normal distribution of activity. Then, limits are fixed in order to determine the degree of deviation from the background activity that will be tolerated before an alert is raised. The single methods listed before realise this general process in various ways.

A *control chart* sets an upper control limit and a lower control limit on the series under monitoring. In surveillance, we are normally interested in getting an alert when a high number of counts appears in the data, i.e. when the upper control limit is exceeded. Assuming a normal distribution, we can estimate the sample mean and standard deviation which are necessary to calculate the upper control limit [73].

The *moving average* algorithm starts from the average of the values in the last $n$ time steps to predict the value for the next time step [31]. To compute an alarm level, a Gaussian distribution is fit to the counts from the recent $n$ days. Again, mean and standard deviation are estimated for the Gaussian. Then, an alarm level can be calculated using the cumulative distribution function of the fitted Gaussian.

*Cumulative sum or CUSUM* methods test to see whether there has been a recent change in the behaviour of time series. It maintains a cumulative sum of deviations from some reference value. The reference value is usually the mean value. Whenever the cumulative sum exceeds a threshold, an alert is raised. The cumulative sum is then set to zero [56]. A signal is triggered when the cumulated differences in observed and expected counts exceed a predetermined threshold [33].

The challenge to be addressed by these approaches is to make a good compromise between detecting true changes and avoiding false alarms. Since these data interpretation methods differ slightly in their underlying assumptions and thus, also in their output and interpretation, the developer of a surveillance system needs to carefully decide for an algorithm. It might be necessary to enable the application of several methods within a surveillance system to allow for multiple interpretations.

Even though these methods are helpful in analysing the data and detected patterns, they do not exactly allow to determine the exact starting date of an event. However, they have been successfully used in surveillance already, mainly in biosurveillance and on structured data such as counts of emergency room visits or numbers of drug prescriptions.

To apply these methods for analysing unstructured data as it is considered here, the data needs to be transformed in some time-series-like format. By applying the text mining methods introduced in the section before, time series can be generated even for previously unstructured data. A time series produced from unstructured data could be for example the frequency of documents over time that contain certain keywords, named entities or met other criteria.

Given the fact that the surveillance data achieved from social sensors might be uncertain and imprecise, it becomes difficult to decide exactly on the existence and nature of an event. Thus, another possibility is to interpret the surveillance data probabilistically. This allows to express the uncertainty about the presence of an event as a probability. When making a decision, this probability can be used in combination with cost and benefit considerations [73]. The statistical methods as presented before are used as detector for events. Then, probabilities can be calculated for the detected event to support the decision process. Possible algorithms are BARD and PANDA that rely upon the Bayes' theorem and have been exploited in Biosurveillance [73].

The field of signal generation and interpretation of surveillance data is still open for research in particular when considering unstructured data as input to the surveillance system. Beyond the methods mentioned before, techniques from anomaly detection could be exploited [59]. These include statistical methods as the ones outlined before, but also machine learning techniques such as classification algorithms.

## 4.5 Ontologies for Surveillance Systems

To perform a semantic analysis of unstructured text, domain knowledge is necessary which is usually provided to surveillance or other knowledge-based systems in form of ontologies.

An ontology is a set of concepts relevant to a particular area of interest, thus, it is extremely domain-dependent. It aims at describing terms and relations of a domain. Ontologies are necessary to represent the relevant prior information of a domain with the objective of adding expressiveness and reasoning capabilities to systems that make use of them.

In the field of medicine lots of different ontologies are already available, e.g. SNOMED, UMLS, GeneOntology. An ontology used within the biosurveillance system BioCaster [13] contains terms such as diseases, agents, symptoms, syndromes. These terms and relations are of course only of interest within the context of biosurveillance. Another surveillance ontology is available for space surveillance. It captures data structure, content, and semantics in the military domain of space surveillance [60].

WordNet [48] is one example for an ontology of general terms, i.e. not domain-specific. It is a so-called top ontology since it comprises general, abstract concepts. In more detail, within WordNet, English nouns, verbs, adjectives and adverbs are organised into sets of synonyms, each representing one underlying lexical concept.

The equivalent to WordNet in German is OpenThesaurus.[11] In December 2011 it contained 80,000 words, and is still growing. It is an interactive website for developing a dictionary of German synonyms. Terms are grouped under categories and generic terms (e.g. category *medicine*, generic term *disease*). OpenThesaurus is also available in other languages such as Polish, Portuguese, Spanish. GeoNames[12] is a graphical database that covers all countries and contains over eight million place-names.

Although there are some ontologies and dictionaries already available, for most application domains, ontologies do not yet exist and often need to be constructed in a time-consuming, manual process. To reduce that effort, approaches have been reported that use domain corpora and clustering algorithms for automatically create ontologies [41]. Another widely used approach is to extract a candidate ontology from a domain corpus and filter it against contrastive corpora, e.g. documents from a general corpus, in order to generate the domain ontology [53]. The contrastive corpora are exploited to select the domain-specific terms.

In surveillance systems, ontologies help to limit the scope of event detection and monitoring to terms and relations necessary to detect events in the domain of interest. An ontology could be used in such system simply as dictionary of domain-specific terms. The number of dictionary terms that can be identified in a document is often hypothesised to provide insights to a potential outcome or to reflect current events. Beyond, ontologies can be of relevance in the various processing steps in a surveillance system, including topic classification, named entity recognition, or event recognition. Also event aggregation can benefit from knowledge in ontologies since they can provide information on categories and other dependencies.

---

[11] http://www.openthesaurus.de

[12] http://www.geonames.org

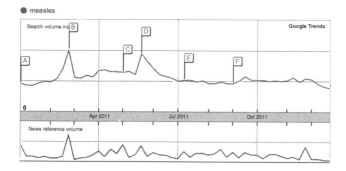

**Fig. 4.5** Visualisation of trends over time as provided by Google trends (http://www.google.com/trends). The trend of search behaviour of the term "measles" in 2011 is shown

## 4.6 Presentation of Information and Alerts

The result of the data analysis and interpretation steps of a surveillance system are signals and related information. This information needs to be communicated to the users. Taking advantage of graphical user interfaces and visualisation capacities of surveillance systems makes it easier for users to investigate detected anomalies. There are mainly two types of systems that can be distinguished and that are characterised by different visualisations: detection systems and information systems.

*Detection systems* are alerting systems. The focus of these systems is not on displaying data for investigation purposes, but on identifying anomalies and patterns and sending alerts. They inform a user about detected patterns and anomalies in form of signals. These systems have normally a static set of visualisation components. If at all, only a minimum of interaction with users is enabled. Typically, detection systems focus on a concrete set of surveillance objects (e.g., diseases, symptoms, topics, opinions) and display only results for objects the system is designed to monitor. Alerts could for example be distributed via E-Mail where no interaction is possible.

Detection results may be shown as line listings, maps, or graphs (e.g., time series graphs). In maps and graphs, information presentation on alerts is highly informative and can be intuitively understood. Table listings allow the user to see all of the alerts on a single view. Nevertheless, users sometimes prefer more aggregated result lists of alerts. Then, they need to have the possibility to drill down to more detailed information.

Surveillance systems that are designed as *information systems* not only provide an alert to a user, but allow to investigate the data underlying the alert and support the human in interpreting the alerts. For this purpose, an information system should be interactive and responsive to user requests, thus allowing users to query and visualise data in a variety of formats. It should incorporate a graphical user interface and multiple data visualisation techniques in order to aid users in deciphering large amounts of data in a timely and cost effective manner. A high level of user interaction is required to enable the user to query and to investigate the results. Additionally, interactions

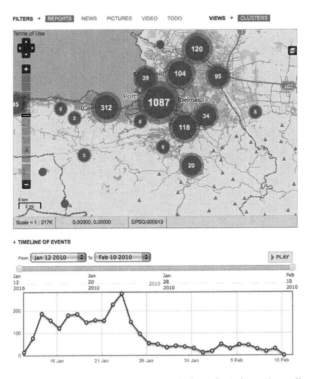

**Fig. 4.6** Visualization of Ushahidi, an open source platform for information collection and inter-active mapping [38] (Image derived from http://www.crunchbase.com/company/ushahidi)

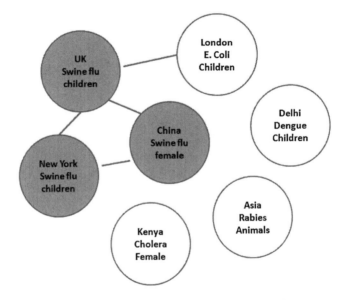

**Fig. 4.7** Visualisation of clustering results. Clusters are associated by links

that allow for future retrieval can be included such as bookmarking, or tagging. Self-learning systems allow also for user rating. Visualisations in information systems comprise time-series graphs, pie charts, maps, tables of data.

Automatically generated time series plots allow for instance users to examine recent trends (monthly, yearly, seasonal) from incoming data streams (see Fig. 4.5).

Geospatial visualisation of data helps users to identify the significance of any recent data anomalies [50]. Geospatial visualisations of detected anomalies can be overlaid with additional information such as infrastructure, hospital locations etc. In general, the mapping layers should include any important infrastructure based on a threat and risk assessment. An example visualisation is shown in Fig. 4.6. It comprises time series plots and interactive mappings of alert levels. Figure 4.7 shows how clustering results can be visualised: the clusters are represented by keywords and they are connected by links.

Visualisation possibilities of events and related information are many-fold. They need to allow a user to easily and quickly find out what is going on or what the alert is about. Visualisations help to assess the risk produced by an event. Again, the user should not be overloaded with information, but should be enabled to dig into the data as much as he or she needs to understand, interpret and react.

# Chapter 5
# An Example: Disease Surveillance from Web 2.0

The biggest efforts with respect to surveillance systems and the largest variety of systems is available for the domain of disease surveillance. In this chapter, the architecture of a possible disease surveillance system is described that exploits Web data as information source. We refer to it as *Hazard Detector*. It serves as an example to demonstrate the nature of event-driven surveillance systems. This system (architecture) is still a vision, but it bases upon a real system. After introducing the problem and task of disease surveillance in Sect. 5.1, the system architecture is described in Sect. 5.2. Some discussions will complete this chapter.

## 5.1 The Task of Disease Surveillance

Various factors such as globalisation, climate change, or behavioural changes contribute to the continuous emergence of public health hazards. A health hazard can be described as a sudden, unexpected event, incident or circumstance, confronting public health officials with a situation threatening the health of people and society with substantial consequences (e.g., the outbreak of an infectious disease such as swine flu or measles). Only the early detection of disease activity followed by an appropriate assessment of its risk and a corresponding reaction can help to reduce and manage the risk produced by such health hazards [58].

Surveillance in public health aims at systematically collecting, analysing, interpreting, and disseminating data about health-related events. The data is used by epidemiologists and public health officials to reduce morbidity and mortality and to improve health. Disease surveillance systems aim at supporting these users in obtaining information on potential health hazards as early as possible. The data on diagnosed cases of notifiable diseases comes traditionally from physicians and laboratories. Since the direct notification can take time, additional sources are used in surveillance systems such as online news or Web 2.0 data [26, 58].

K. Denecke, *Event-Driven Surveillance*, SpringerBriefs in Computer Science,
DOI: 10.1007/978-3-642-28135-8_5, © The Author(s) 2012

From a user perspective, a disease surveillance system needs to consider five main issues to be useful:

1. **Content collection**. The system should monitor a broad range of sources in a multitude of languages and from being broadcasted or produced around the globe. Besides monitoring diseases and their mentions, it is of interest to monitor symptoms and their mentions as well as behavioural changes.
2. **Alert generation**. The system should produce signals or alerts that notify the user when a health event is detected. For a signal produced, information is provided to allow the user an in-depth analysis of the detected event.
3. **Result filtering**. Results, i.e. signals need to be carefully filtered according to various filter criteria such as relevance, novelty, source of information etc. to avoid information overload.
4. **Result presentation**. Event information should be presented in a structured, appropriate and user-friendly way and should allow accessing the original information sources for event verification processes. The user need to interact with the results, e.g., by narrowing the result set or by redefining the user interest.
5. **User feedback and interaction**. User interactions with a disease surveillance system should include (a) the specification of essential characteristics of events of interest (e.g., disease name, location), and (b) a selection of result presentation formats, (c) the possibility of storing event information and (d) the possibility of providing feedback for future result adaptations.

In the following, a typical scenario for disease surveillance is described highlighting the main interactions between events, system and user (see Fig. 5.1).

---

**Scenario: Disease surveillance from Web 2.0**


The user of our surveillance system is supposed to be an expert for vaccine-preventable infections at a local health organisation, and has a user profile to signals related to measles, mumps and rubella (MMR), etc. He regularly checks for signals produced by the system that are related to his interests. One day, a group of pupils is tweeting about several class mates being absent from school because of fever, rash etc. There is speculation about measles. The surveillance system identifies to what degree these messages match respective keywords, and extracts essential event information (e.g. age, location, student-status). Indicators as, for instance, the number of documents matching specified keywords are identified. At some point, the system generates a signal based on the extracted indicator information and communicates this signal to the user because the analysis indicates an unusual event. Since this signal refers to events that fall into the field of interest of the user, it is presented to him. To validate the signal received, the user checks his department's data and calls the local public health department for getting further information or implementation of control measures.

---

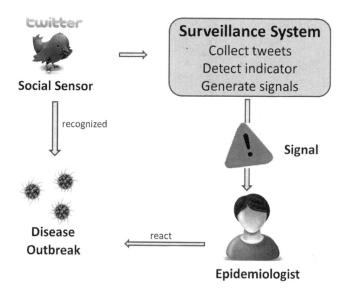

**Fig. 5.1** Scenario: disease surveillance from Web 2.0

## 5.2 Event-Driven Architecture for Disease Surveillance

The system architecture of the *Hazard Detector* consists of various technical components that realise the individual processing steps. Its characteristics include:

- Text stream collection. Text streams are collected from various data sources (e.g., online news, blogs, twitter, health indicator data [23]);
- Text stream analysis. Text streams are processed among others by data cleansing, named entity recognition, filtering technologies;
- Identification of patterns of interest. Pattern among texts that provide indications to public health events are identified;
- Notification of systems and personnel. An alert is sent to those systems or personnel that specified their interest.

Figure 5.2 shows the *Hazard Detector* system architecture. It can be seen as a client/server architecture, where the client connects to the user and the server is responsible for triggering the single components by forwarding event messages. Beyond, a set of external tools and knowledge sources are exploited by the system.

In general, the system works as follows. Knowing the user interest, it continuously monitors the Web for incoming text and data streams related to relevant health events. Once texts of interest are identified, appropriate services for pattern analysis and interpretation are triggered one after another by forwarding system event messages. At the end of the pipeline, an alert is produced when certain conditions are fulfilled. The single components will be described in the following.

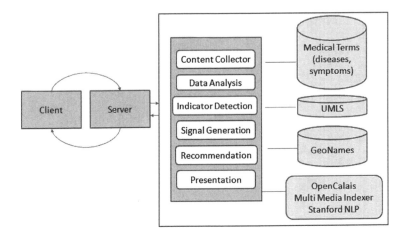

**Fig. 5.2** Hazard detector system architecture

## 5.2.1 Content Collection Component

This component fills continuously the information database of the system. It collects data from social sensors and other sources, and makes them accessible to the processing components. The content collection focuses on two groups of data:

- online news data, and
- data from social sensors like blogs, forums and Twitter.

The content collection component actively collects data from the various sensors or sources. It sends keyword searches to retrieve documents matching these keywords. Among others, it collects regularly RSS feeds provided by online news channels or news monitoring channels such as MedISys [66]. MediSys is a monitoring system that checks online news for disease-related keywords. These online news are provided along with additional annotations among others on the identified locations.

Data from social sensors is collected by crawling blogs and forum pages that are registered in the system. This enables users to add sources they are interested in monitoring.

Additionally, data collection tools and APIs are used. The Twitter API is exploited to get tweets matching specific, predefined keywords. Additional blog data is collected using Spinn3R. The search terms required by Spinn3R are specified by domain experts and comprise disease names and terms describing medical symptoms.

Once a specific amount of data is available a system event is produced that triggers the document analysis component via the server.

## 5.2.2 Document Analysis Component

The *document analysis component* filters and pre-processes the collected (textual) data before making it available for the event detection component. Pre-processing includes filtering of irrelevant data, recognition of mentions of disease names and symptoms, locations, time etc. by means of existing text processing tools (e.g., Open-Calais[1] and Minipar[2]). As a result, a set of documents annotated with named entities, parts of speech and linguistic structures is produced. These tagged documents are indexed and made available for further processing. Once the document analysis is completed for the available data, a system event is sent to the server that triggers the next component.

## 5.2.3 Event Detection and Signal Generation Component

This component comprises two main steps: first, from the data provided by the document analysis component on the collected documents indicators are detected. With the term *indicator* we mean in this context a set of documents or a set of sentences that might be an indication to the occurrence of a disease outbreak. Second, detected indicators are used to generate signals or alerts to be provided to the user.

The event detection component identifies from the annotated documents relevant sentences or subsections. A sentence (subsection) is considered relevant, when it reports information on a health event (referred to as disease-reporting sentences or indicators) [67]. The reason for this filtering is that in long documents only some sentences or paragraphs may be relevant for health event detection. In particular in blogs, people are changing topics, writing about the weather, sickness and someones birthday in one posting.

The relevant sections are represented further by disease-location pairs. Using this representation, time series are generated and biosurveillance algorithms for signal generation are applied such as RKI-method, Stroup's method or CUSUM [36]. They allow analysing the data and generating a signal when certain thresholds are passed. Once a signal is generated, a system event is produced and the recommendation component is triggered.

## 5.2.4 Recommendation Component

The recommendation component gets as input the signals and related texts calculated from the previous component. Generated signals and related information are matched by the recommendation component with the signal definition entered by the user for selecting relevant signals. It ranks the signals and related document appropriately

---

[1] http://www.opencalais.com

[2] http://webdocs.cs.ualberta.ca/lindek/minipar.htm

and takes additionally the user profile into account. The user profile consists of information on specified signal definitions as well as user feedback from previous searches provided through a tagging and rating facility in the user interface.

The user profile is used to build a user and recommendation model. Based on that model, signals that might be of potential interest to the user are selected from the generated signals. The idea behind is that a user seldom knows in advance which health events might occur. Thus, formulating a query for a signal search is difficult and relevant signals might be missed. For this reason, the recommendation component is intended to select signals that are of interest for the single user.

Additionally, overwhelming the user with irrelevant signals or indicators should be avoided. The user is asked for rating signals and indicators. Based on the rating, the weighting of indicators for signal generation can be adapted to filter or rerank the presented indicators (i.e. documents) or signals. Once the recommendation component completed re-ranking, again a system event is generated letting the presentation component know that there is new information to present in the interface.

### 5.2.5  User Interface

The user interface (see Figs. 5.3, 5.4) allows a user to specify his interest in terms of a signal definition and to investigate the signals produced by the system. The signal definition specifies the object under monitoring. It comprises one or more disease names or symptoms, locations to be considered as well as sources to be checked by the surveillance system.

Once signals are generated that match a signal definition (i.e. they are related to the surveillance task), they are presented to the user. For instance, the documents from which a signal was generated and the information source are presented to the user in a graphical user interface. Users are enabled to browse through the results. Various visualisation methods are applied to present the results in an easily understandable way (e.g., as word clouds, in maps or graphs). Beyond, the user interface supports interacting and commenting of signals, tagging and rating of signals and related documents.

Tagging and rating facilities are necessary to get additional data on the user interest. It allows the user to add tags to signals and to indicate whether a signal or document was relevant or not. The components, mainly the recommendation, exploit this information for future result presentations.

## 5.3  Experiences and Perspectives

The system architecture as it was described in this chapter does not yet exist even though a similar system is developed in the M-Eco project[3] [18]. However, the architecture of the M-Eco system is not necessarily following the event-driven architecture

---

[3] http://www.meco-project.eu

## Hazard Detector - Signal Definition

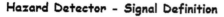

| Diseases and Symptoms | Region | Source |
|---|---|---|
| Cholera | Africa | TV |
| H1N1 | America | Radio |
| Influenza (A,B,C) | Asia | Blog |
| Malaria | Australia | Microblogs |
| Measles | Europe | Forum |
| Rabies | | News |
| | Albania | |
| Cough | Andorra | |
| Fever | Austria | |
| Headache | Belgium | |
| Sore throat | Bosnia | |
| Pain | Bulgaria | |

**Fig. 5.3** Signal definition page of the hazard detector system

#### Hazard Detector

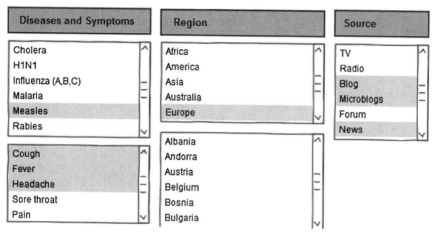

**Recent Alerts**

| Disease | Location | Time |
|---|---|---|
| Chikungunya | India | 03.12.2011 |
| Measles | Germany | 12.10.2011 |
| EHEC | Germany, Hamburg | 30.05.2011 |

**Associated Documents**

Chikungunya is yet to subside in the city despite the onset of . winter. We have currently 70 new cases. [...]

India: Dengue, chikungunya and encephalitis continue in New Delhi

Another mosquito-borne disease, chikungunya, too continues to be reported from different hospitals in Delhi.

**Fig. 5.4** Result presentation of the hazard detector system

paradigm. This chapter is finished by a discussion of benefits and shortcomings of using the principle of an event-driven architecture for such disease surveillance system. The components of the current implementation of the M-Eco system (state December 2011) are not triggered by events, but check regularly for updates in the database. To demonstrate an event-driven architecture, the described architecture is a modified version of that system. The benefit of implementing such system as event-driven architecture is timeliness: the system can faster detect changes or anomalous behaviour, when it is event-driven. The M-Eco system as it is currently designed processes the data every night and is thus late in generating alerts in critical situations.

A problem is that analysing unstructured texts by natural language processing tools can take some time depending on the amount of data to be processed. For this reason, the M-Eco system is currently not implemented as event-driven architecture: the performance of the underlying tools and components is not yet sufficient.

An experience made in M-Eco is that it is crucial to carefully select the data sources to be monitored and to use appropriate methods for ensuring that all relevant data can be considered by the systems. Given the amount of social media data available, it is necessary to collect based on keywords. This limits the data collection to data that matches the keywords. A good approach to address this limitation is to enable an easy adaptation of the keywords. This can be done by manual entry to some maintenance interface.

# Chapter 6
# Future Challenges

Surveillance of situations, subjects or objects is crucial in various scenarios. Even though we are not threaten by lions or missing water holes like animals on the savannah, there are many situations and threats that require active monitoring. Modern society accompanied with terrorism, natural disasters and others, make these threats even worse. Maybe in future, I will have an App on my mobile that provides information where to go when a serious flood happened, just by skimming through the Web chatter and aggregating the information available.

The preceding chapters highlighted the technological opportunities and possibilities of surveillance from unstructured texts, mainly from Web data. Exploiting information exchanged through the Web as information source for monitoring just began. However, it is already clear that one of the greatest assets of future surveillance systems that use data from social sensors is their timeliness. These systems receive data in real-time or near real-time, and thus provide the opportunity to detect situational or behavioural changes faster than through traditional reporting methods or other sensors. The biggest challenges are the huge amounts of data and the unstructured format, the natural language as such which makes automatic analysis difficult due to its rich variety. As we have seen, research and technology developed already useful tools to support the automatic analysis of written natural language.

Although surveillance from Web data is already possible from the technological point of view, there are several open issues to be addressed in future. In surveillance systems that are designed to support risk assessment, it is crucial to provide as much information as possible about a detected event. For this reason, event enrichment is necessary, i.e. information from heterogeneous sources need to be combined to provide events of higher level abstraction. It is often not sufficient to only discover that something happened or that suddenly lots of documents mention a specific keyword. The user is interested in seeing the complete picture and in learning about the origin of an event. What did the system recognised and led to the generation of an alert? This question need to be answered by the system to allow its users to assess the relevance of the alert and its associated risk. Therefore, additional metadata need

K. Denecke, *Event-Driven Surveillance*, SpringerBriefs in Computer Science, 63
DOI: 10.1007/978-3-642-28135-8_6, © The Author(s) 2012

to be automatically extracted from unstructured data in order to give the user the opportunity to investigate a detected event and related information.

In this context it is also a future challenge to aggregate events detected by different sensors. The specific challenge here is that sensors are providing data in differing update frequencies. The data is not always directly comparable or extendable. Just consider a surveillance system that relies upon sensors that provide structured and unstructured data. The different inputs need to be processed and prepared in a way that they can be somehow combined. How can data from different sources be linked? This question still needs to be addressed.

Another open issue for future research is to ensure interoperability of surveillance systems. In some domains, surveillance systems are already available. We have seen that existing biosurveillance systems exploit different sources. From the user perspective, it is unrealistic to check with several systems when searching for an event. However, the systems can provide similar information, or can complement each other. There has not yet been an emphasis on the automated exchange of surveillance results between systems. Some surveillance systems allow for doing surveillance in a restricted region. For getting a more comprehensive overview on a larger area and for allowing also for re-use of detected events in other contexts, integration of data provided by several systems is crucial. From a user perspective, such integration would also make life easier since in time critical situations there is simply no time to access several complementing surveillance systems.

A possible solution for common data delivery would be to let surveillance algorithms provide data (structured data) as RSS or via Web services. This data could then be combined in a common user interface. One effort to mention here is the National Retail Data Monitor [69] which provides the result data through Web services. Another solution might be the specification of a record of events of significance, that are stored, analysed, tracked and exchanged between surveillance systems. This could help to increase the coverage and effectiveness of surveillance as well.

Besides integration of complementing surveillance systems or their results, development towards flexible tools would be beneficial. We envision a framework that sets up automatically a surveillance system for a concrete monitoring task for which a user specifies some parameters such as:

- a set of data sources,
- an update rate for content collection,
- extractors for finding entities and events,
- a taxonomy to specify the domain.

With this minimum of information, a surveillance system for a concrete task should be easily be created.

The signal generation methods that are already available mainly rely upon statistics. Sometimes more sophisticated reasoning might be necessary to decide whether detected events need to be interpreted as signals. Thus, a significant challenge for the future is the construction of reasoning engines to process events. Reasoning could also support in making predictions on future events based on an retrospective analysis and consideration of the current situations.

A problem with event detection methods is that they suffer when different vocabularies are used in the texts under consideration. Given the evolution of language and different styles of writing and expressing things, these methods are difficulty to apply successfully on online texts. Systems need to be enabled to actively learn new vocabulary.

Research-wise evaluation of event detection algorithms and surveillance systems is an interesting open issue. Missing standard evaluation corpora avoid the comparison of the quality of existing approaches. Often, a comparison is even impossible due to different surveillance tasks or domains that the systems base on. The problem here is that so many different things can be under monitoring that the systems differ significantly. At the end, it is of course the user who decides about the usefulness and reliability of system's outputs. Thus, in all evaluation efforts, interaction with users is necessary to continuously improve technology, vocabulary and information sources.

Open source data is the most timely, but also the most unreliable of information. There is always the risk of getting spammed by some content providers. In such a case, a surveillance system would create a false alarm. However, even though there is a risk of getting false alarms, the benefits still have a higher weight. Additional work is needed to standardise the assessment of content reliability and spam detection for better evaluating detected events [30].

Accompanied with surveillance and monitoring, we should start to consider ethical issues and privacy concerns. Is monitoring well accepted as long as the single individual is not concerned by it? Or do we agree in being surveilled to be safe in life threatening situations? The process of collecting and aggregating data on person's behaviour is aimed at the production of a "cyber-portraiture" or a "data shadow" which exists for nearly everyone and calls into question the notion of informational self-determination. Clearly, in disastrous situations or life threatening situations almost everyone would agree (and probably also contribute) in getting monitored. But in "normal life"? When setting up a surveillance system, people need to consciously agree both in providing their personal data and in being tracked by social network services. On the other hand, new rules for respecting and preserving the user privacy have to be formulated. Otherwise, we risk the misuse of such data.

In conclusion, technology allows to build sophisticated surveillance systems that support humans in being informed early on changing situations. There is an ongoing need and interest in having surveillance system in multiple domains and application areas. However, there are still challenges to address, which should be encouraging for researchers to continue finding solutions to them.

# Glossary

| | |
|---|---|
| **Alert** | An alert or signal is created when an event of relevance is detected, for example when the number of daily counts exceeds some threshold. |
| **Event** | An event is a significant change in the state of a situation, environment or person. Whether a change is significant depends on the user. |
| **Event-driven Architecture, EDA** | An EDA is a type of software architecture in which components are driven by the occurrence of events and communicate by means of events. |
| **False positive** | A false positive is defined as an alert that is raised in the absence of a real world event |
| **Indicator** | An indicator is a hint to an event. For example the number of documents mentioning some specific term is an indicator. |
| **Monitoring** | Monitoring is the active or passive routine collection of information on a object or process with the purpose of detecting changes over time. |
| **Natural language processing (NLP)** | Natural language processing encompasses a range of computational techniques for analysing and representing natural language texts. They target at making unstructured data available for a range of tasks or applications. |
| **Ontology** | An ontology is a set of concepts relevant to a particular area of interest. |

K. Denecke, *Event-Driven Surveillance*, SpringerBriefs in Computer Science, DOI: 10.1007/978-3-642-28135-8, © The Author(s) 2012

| | |
|---|---|
| **Part of speech (POS)** | A POS is the linguistic category of a word. |
| **Part of speech tagger** | A POS tagger is a tool that assigns a word class to a word. |
| **Parsing** | Parsing is the process of breaking down a sentence into its constituents parts. A full parser connects all words and phrases together into a sentence while a shallow parser combines words into noun phrases, verb phrases and prepositional phrases, but resists on linking the phrases together. |
| **RSS** | Really Simple Syndication is a family of web feed formats used to publish frequently updated works. |
| **Signal** | See Alert. |
| **Situational awareness** | Situational awareness is the ability to identify, process, and comprehend the critical elements of information about what is happening to the team with regards to the mission. |
| **Social text stream** | Social text streams are defined as collection of informal text communication distributed over theWeb. Each piece of text is associated with social attributes such as author. |
| **Social sensor** | A social sensor is any source of information that can be identified in Web 2.0 tools. In such sources, situations and facts about users are expressed either by the user himself, by others or just through his interaction with these tools. Examples of social sensors include Twitter posts, Facebook status updates or pictures posted on Flickr. |
| **Stream** | To a set of continuously arriving events we refer as a stream. |
| **Sublanguage** | The informational content and structure of a domain form a specialised language called sublanguage. |
| **Surveillance** | Surveillance is the systematic, continuous monitoring of objects, persons or situations. |
| **Surveillance system** | Surveillance systems monitor, track and assess the movements of individuals, their property and other assets. |
| **Time series** | A time series is a sequence of observations made over time. |

**Transfer learning**

Transfer learning or inductive learning is a research problem in machine learning that focuses on storing knowledge gained while solving one problem and applying it to a different but related problem.

**Web mining**

Web mining targets at finding patterns in Web data. It comprises the tasksWeb usage mining,Web content mining and Web structure mining.

# References

1. E. Agichtein, L. Gravano, J. Pavel, V. Sokolova, A. Voskoboynik, Snowball, a prototype system for extracting relations from large text collections. SIGMOD Rec. **30**, 612 (2001)
2. S. Ahsan, A. Shah, Data mining, semantic web and advanced infromation technologies for fighting terrorism, in *Proceedings of IEEE International Sympo- sium on Biometrics and Security Technologies (ISBAST 2007)*
3. J. Allan, V. Lavrenko, R. Papka, Event tracking. IR IR-128. University of Massachusetts (1998)
4. J. Allan, R. Papka, V. Lavrenko, Online new event detection and tracking, in *Proceedings of 21st Annual International ACM SIGIR Conference on Research and Development in Information Retrieval*, pp. 37–45 (1998)
5. B. Babcock, S. Babu, M. Datar, R. Motwani, J. Widom, Models and issues in data stream systems, in L. Popa (ed.) *PODS*, pp. 1–16. ACM (2002)
6. A. E. S. Baccianella, F. Sebastiani, Sentiwordnet 3.0: An enhanced lexical resource for sentiment analysis and opinion mining, in *Proceedings of the Seventh conference on International Language Resources and Evaluation (LREC'10)* (European Language Resources Association (ELRA), Valletta, 2010)
7. A. Bifet, R. Kirkby, Data stream mining: a practical approach. Technical report, The University of Waikato, Aug 2009
8. E. Boiy, P. Hens, K. Deschacht, M.-F. Moens, Automatic sentiment analysis in on-line text, in *Proceedings of the 11th International Conference on Electronic Publishing held in Vienna, Austria 13–15 June 2007*, pp. 349–360 (2007)
9. R. Bruns, J. Dunkel, *Event-driven architecture: Softwarearchitektur fr ereignisgesteuerte Geschftsprozesse* (Springer, Berlin, 2010)
10. C. Castillo, *Web crawling*, Chapter 12 (2010)
11. E.H. Chan, V. Sahai, C. Conrad, J.S. Brownstein, Using web search query data to monitor dengue epidemics: a new model for neglected tropical disease surveillance. *PLoS Negl. Tropical Dis.*, **5**(5):e1206+ (2011)
12. C.-H. Chang, M. Kayed, M.R. Girgis, K.F. Shaalan, A survey of web information extraction systems. *IEEE Trans. Knowl. Data Eng.*, **18**:1411–1428 (2006)
13. N. Collier, R.M. Goodwin, J. McCrae, S. Doan, A. Kawazoe, M. Conway, A. Kawtrakul, K. Takeuchi, D. Dien, An ontology-driven system for detecting global health events, in *Proceedings of the 23rd International Conference on Computational Linguistics*, COLING '10, Stroudsburg, PA, USA, pp. 215–222 (2010). Association for Computational Linguistics
14. J. Cowie, W. Lehnert, Information extraction. Commun. ACM **39**, 80–91 (1996)
15. V. Cristea, F. Pop, C. Dobre, A. Costan, *Distributed Architectures for Event-Based Systems*. Studies in Computational Intelligence, vol. 347, chapter 2 (Springer, Berlin, 2011), pp. 11–45

16. A. Culotta, M. Wick, R. Hall, A. Mccallum, First-order probabilistic models for coreference resolution, in *Proceedings of HLT-NAACL 2007*
17. S.R. Das, M.Y. Chen, Yahoo! for amazon: Sentiment extraction from small talk on the web. Manage Sci **53**(9), 1375–1388 (2007)
18. K. Denecke, G. Kirchner, P. Dolog, P. Smrz, J. Linge, G. Backfried, J. Dreesman, Event-driven architecture for health event detection from multiple sources, in *Proceedings of the XXIII International Conference of the European Federation for Medical Informatics (MIE 2011)*, 28–31.August, 2011, Oslo, Norway (2011)
19. M. DeRosa, Data mining and data analysis for counter terrorism (2004)
20. N.A. Diakopoulos, D.A. Shamma, Characterizing debate performance via aggregated twitter sentiment, in *Proceedings of the 28th International Conference on Human Factors in Computing Systems*, CHI '10 (ACM, New York, 2010), pp. 1195–1198
21. S. Dumais, J. Platt, M. Sahami, D. Heckerman, Inductive learning algorithms and representations for text categorization (ACM Press, New York, 1998), pp. 148–155
22. O. Etzioni, M. Banko, S. Soderland, D.S. Weld, Open information extraction from the web. Commun. ACM **51**, 68–74 (2008)
23. D. Faensen, H. Claus, J. Benzler, Survnet@rki—a multistate electronic reporting system for communicable diseases. Euro Surveillance **11**(4), 100–103 (2006)
24. R. Feldman, J. Sanger, *The Textmining Handbook* (Cambridge University Press, Cambridge, 2007)
25. M. Fisichella, A. Stewart, K. Denecke, W. Nejdl, Unsupervised public health event detection for epidemic intelligence, in *CIKM*, ed. by J. Huang, N. Koudas, G.J.F. Jones, X. Wu, K. Collins-Thompson, A. An (ACM, New York, 2010), pp. 1881–1884
26. C.F. Freifeld, K.D. Mandl, B.Y. Reis, J.S. Brownstein, Healthmap: global infectious disease monitoring through automated classification and visualization of internet media reports. J. Am. Med. Inform. Assoc. **15**, 150–157 (2008)
27. C. Friedman, P.K.A. Rzhetsky, Two biomedical sublanguages: a description based on theories of zellig harris. J. Biomed. Inform. **35**, 222–235 (2002)
28. M.M. Gaber, A. Zaslavsky, S. Krishnaswamy, Mining data streams: a review. SIGMOD Rec. **34**, 18–26 (2005)
29. J. Ginsberg, M.H. Mohebbi, R.S. Patel, L. Brammer, M.S. Smolinski, L. Brilliant, Detecting influenza epidemics using search engine query data, Nature **457**(7232), 1012–1014 (2008)
30. N. Grady, L.Vizenor, J. Marin, L. Peitersen, Bio-surveillance event models, open source intelligence, and the semantic web, in *BioSecure*, ed by D. Zeng et al. LNCS, vol. 5354 (2008), pp. 22–31
31. J. Hamilton, *Time Series Analysis* (Princeton UNiversity Press, Princeton, 1994)
32. J. Hanson, Event-driven services in soa. *Javaworld* (2005)
33. D. Hawkins, D. Olwell, *Cumulative Sum Charts and Charting for Quality Improvement* (Springer, New York, 1998)
34. H. Becker, M. Naaman, L. Gravano, Beyond trending topics: real-world event identification on twitter, in *Proceedings of the Fifth International AAAI Conference on Weblogs and Social Media* (2011), pp. 438–441
35. F. Hogenboom, F. Frasincar, U. Kaymak, F. de Jong, An overview of event extraction from text, in *Workshop on Detection, Representation, and Exploitation of Events in the Semantic Web (DeRiVE 2011) at Tenth International Semantic Web Conference (ISWC 2011)*, ed by M. van Erp, W. R. van Hage, L. Hollink, A. Jameson, R. Troncy. CEUR Workshop Proceedings, vol. 779, CEUR-WS.org (2011), pp. 48–57
36. M. Höhle, Surveillance: An R package for the surveillance of infectious diseases. Comput. Stat. **22**(4), 571–582 (2007)
37. A.L. Hughes, L. Palen, J. Sutton, S. liu, S. Vieweg, Site-seeing in disaster: an examination of on-line social convergence, in *2008 ISCRAM Conference* (2008)

38. F. Johansson, J. Brynielsson, P. Hrling, M. Malm, C. Mrtenson, S. Truv, M. Rosell, Detecting emergent conflicts through web mining and visualization, in *EISIC* (IEEE, 2011), pp. 346–353

39. D. Jurafsky J.H. Martin, *Speech and Language Processing (2nd Edition) (Prentice Hall Series in Arti_cial Intelligence)*, 2nd edn. (Prentice Hall, Upper Saddle River, 2008)

40. N. Kambhatla, Combining lexical, syntactic, and semantic features with maximum entropy models for extracting relations, in *Proceedings of the ACL 2004 on Interactive Poster and Demonstration Sessions*, ACLdemo '04, Stroudsburg, PA, USA (Association for Computational Linguistics, 2004)

41. L. Khan, F. Luo, Ontology construction for information selection, in *ICTA*. (IEEE Computer Society, 2002), p. 122

42. J. Kleinberg, Bursty and hierarchical structure in streams. Data Min. Knowl. Discov. **7**, 373–397 (2003)

43. G. Kumaran, J. Allan, Text classification and named entities for new event detection, in *SIGIR* (2004), pp. 297–304

44. H. Lee, Y. Peirsman, A. Chang, N. Chambers, M. Surdeanu, D. Jurafsky, Stanford's multi-pass sieve coreference resolution system at the conll-2011 shared task, in *Proceedings of the CoNLL-2011 Shared Task* (2011)

45. J. Lombardo, D. Buckeridge, *Disease Surveillance: A Public Health Informatics Approach* (Wiley, 2007)

46. D. Luckham, *The Power of Events: An Introduction to Complex Event Processing in Distributed Enterprise Systems* (Addison-Wesley Professional, 2002)

47. C. Manning, H. Schutze, *Foundations of Statistical Natural Language Processing*. MIT Press, Cambridge (1999)

48. G.A. Miller, Wordnet: a lexical database for english. Commun. ACM **38**, 39–41 (1995)

49. R. Mitkov, *The Oxford Handbook of Computational Linguistics*, Oxford University Press, New York (2005)

50. K.M. Moore, G. Edge, A.R. Kurc, Visualization techniques and graphical user interfaces in syndromic surveillance systems. Summary from the disease surveillance workshop, 11–12, Sep, 2007, Bangkok, Thailand, in *BMC Proc.*, Suppl 3(2) (2008)

51. T. Mullen, R. Malouf, A preliminary investigation into sentiment analysis of informal political discourse, in *AAAI 2006 Spring Symposium on Computational Approaches to Analysing Weblogs (AAAI-CAAW 2006)*

52. M. Naughton, N. Stokes, J. Carthy, Sentence-level event classification in unstructured texts. *Information Retrieval* (2009)

53. R. Navigli, P. Velardi, Automatic adaptation of wordnet to domains, in *Proceedings of 3rd International Conference on Language Resources and Evaluation*, Las Palmas, Canary Island, Spain (2002)

54. V. Ng, C. Cardie, Improving machine learning approaches to coreference resolution (2002)

55. D.O. Olguin, A. Pentland, Social sensors for automatic data collection, in *Proceedings of the Fourteenth Americas Conference on Information Systems*, Aug 2008

56. E.S. Page, Continuous inspection schemes. Biometrika **41**(1/2), 100 (1954)

57. B. Pang, L. Lee, Opinion mining and sentiment analysis. Found. Trends Inf. Retr. **2**, 1–135 (2008)

58. C. Paquet, D. Coulombier, R. Kaier, M. Ciotti, Epidemic intelligence: a new framework for strengthening disease surveillance in Europe. *Euro Surveillance* **11**(12) (2006)

59. A. Patcha, J.-M. Park, An overview of anomaly detection techniques: existing solutions and latest technological trends. Comput. Netw. **51**, 3448–3470 (2007)

60. M. Pulvermacher, D. Brandsma, J. Wilson, A space surveillance ontology: captured in an xml schema (2006)

61. K. Roberts, S.M. Harabagiu, Detecting new and emerging events from textual sources, in *Fifth IEEE International Conference on Semantic Computing* (2011), pp. 67–74

62. A. Rosi, S. Dobson, M. Mamei, G. Stevenson, J. Ye, F. Zambonelli, Social sensors and pervasive services: approaches and perspectives, in *PerCol2011: Proceedings of the 2nd International Workshop on Pervasive Collaboration and Social Networking*, March 2011

63. T. Sakaki, M. Okazaki, Y. Matsuo, Earthquake shakes twitter users: realtime event detection by social sensors, in *WWW*, ed by M. Rappa, P. Jones, J. Freire, S. Chakrabarti (ACM, 2010), pp. 851–860

64. S. Soderland, Learning information extraction rules for semi-structured and free text, Mach. Learn. **34**(1–3), 233–272 (1999)

65. K. Starbird, L. Palen, A.L. Hughes, S. Vieweg, Chatter on the red: what hazards threat reveals about the social life of microblogged information, in *Proceedings of the CSCW Conference* (2010)

66. R. Steinberger, F. Fuart, E. van der Goot, C. Best, P. von Etter, R. Yangarber, Text mining from the web for medical intelligence, in *Mining Massive Data Sets for Security*, ed by F.-S. Françoise et al. (IOS Press, 2008)

67. A. Stewart, K. Denecke, Using promed mail and medworm blogs for crossdomain pattern analysis in epidemic intelligence, in *Studies in Health Technology and Informatics: MEDINFO 2010*, ed by C. Safran, S. Reti, H. Marin (IOS Press, 2010), pp. 473–481

68. P. Tetlock, Giving content to investor sentiment: the role of media in the stock market, J. Finance **62**, 1139–1168 (2007)

69. F. Tsui, J. Espino, Y. Weng, A. Choudary, H. Su, M. Wagner, Key design elements of a data utility for national biosurveillance: event-driven architecture, caching, and web service model, in *AMIA Annu Symp Proc.* (2005), pp. 739–743

70. R. Valitutti, Wordnet-affect: an affective extension of wordnet, in *Proceedings of the 4th International Conference on Language Resources and Evaluation* (2004), pp. 1083–1086

71. P.K. Varshney, I.L. Coman, Issues and recent advances in distributed multi-sensor surveillance, in *Proceedings of the 2nd European Workshop on Advanced Video-Based Surveillance Systems 2001, Distributed Surveillance* (2001), pp. 41–52

72. S. Vieweg, A.L. Hughes, K. Starbird, L. Palen, Microblogging during two natural hazards events: what twitter may contribute to situational awareness, in *Proceedings of the CHI Conference 2010*, Atlanta (2010), pp. 1079–1088

73. M. Wagner, A.W. Moore, R.M. Aryel, *Handbook of Biosurveillance* (Elsevier Academic Press, Oxford, 2006)

74. C. Wei, Y. Lee, Event detection from online news documents for supporting environmental scanning. Decision Support Syst. **36**, 385–401 (2004)

75. I.H. Witten, E. Frank, *Data Mining: Practical Machine Learning Tools and Techniques*, The Morgan Kaufmann Series in Data Management Systems, 2nd edn (Morgan Kaufmann Publishers, San Francisco, 2005)

76. Q. Yang, An introduction to transfer learning, in *Proceedings of the 4th International Conference on Advanced Data Mining and Applications*, ADMA '08 (Springer-Verlag, Berlin, 2008), p. 1

77. Y. Yang, J. Carbonell, R. Brown et al., Learning approaches for detecting and tracking news events. IEEE Intell. Syst. **14**(4), 32–43 (1999)

78. Q. Zhao, P. Mitra, Event detection and visualization for social text streams, in *ICWSM 2007 Boulder Canada* (2007)

79. L. Zhiwei, L. Mingjing, M. Weiying, A probabilistic model for retrospective news event detection, in *Proceedings of SIGIR 05* (2005), pp. 106–113

# Index

Printed by Publishers' Graphics LLC USA
MO20120327-131
2012